COMPUTERS AND GRAPH THEORY
Representing Knowledge for Processing by Computers

ELLIS HORWOOD SERIES IN COMPUTERS AND THEIR APPLICATIONS

Series Editor: IAN CHIVERS, Senior Analyst, The Computer Centre, King's College, London, and formerly Senior Programmer and Analyst, Imperial College of Science and Technology, University of London

Abramsky, S. & Hankin, C.J.	ABSTRACT INTERPRETATION OF DECLARATIVE LANGUAGES
Alexander, H.	FORMALLY-BASED TOOLS AND TECHNIQUES FOR HUMAN–COMPUTER DIALOGUES
Atherton, R.	STRUCTURED PROGRAMMING WITH BBC BASIC
Atherton, R.	STRUCTURED PROGRAMMING WITH COMAL
Baeza-Yates, R.A.	TEXT SEARCHING ALGORITHMS
Bailey, R.	FUNCTIONAL PROGRAMMING WITH HOPE
Barrett, R., Ramsay, A. & Sloman, A.	POP-11
Beardon, C., Lumsden, D. & Holmes, G.	NATURAL LANGUAGE AND COMPUTATIONAL LINGUISTICS: An Introduction
Berztiss, A.	PROGRAMMING WITH GENERATORS
Bharath, R.	COMPUTERS AND GRAPH THEORY: Representing Knowledge for Processing by Computers
Bishop, P.	FIFTH GENERATION COMPUTERS
Bullinger, H.-J. & Gunzenhauser, H.	SOFTWARE ERGONOMICS
Burns, A.	NEW INFORMATION TECHNOLOGY
Carberry, J.C.	COBOL
Carlini, U. & Villano, U.	TRANSPUTERS AND PARALLEL ARCHITECTURES
Chivers, I.D.	AN INTRODUCTION TO STANDARD PASCAL
Chivers, I.D.	MODULA 2
Chivers, I.D. & Sleighthome, J.	INTERACTIVE FORTRAN 77
Clark, M.W.	PC-PORTABLE FORTRAN
Clark, M.W.	TEX
Cockshott, W. P.	A COMPILER WRITER'S TOOLBOX: How to Implement Interactive Compilers for PCs with Turbo Pascal
Cockshott, W. P.	PS-ALGOL IMPLEMENTATIONS: Applications in Persistent Object-Oriented Programming
Colomb, R.	IMPLEMENTING PERSISTENT PROLOG
Cope, T.	COMPUTING USING BASIC
Curth, M.A. & Edelmann, H.	APL
Dahlstrand, I.	SOFTWARE PORTABILITY AND STANDARDS
Dandamudi, S. P.	HIERARCHICAL HYPERCUBE MULTICOMPUTER INTERCONNECTION NETWORKS
Dongarra, J., Duff, I., Gaffney, P., & McKee, S.	VECTOR AND PARALLEL COMPUTING
Dunne, P.E.	COMPUTABILITY THEORY: Concepts and Applications
Eastlake, J.J.	A STRUCTURED APPROACH TO COMPUTER STRATEGY
Eisenbach, S.	FUNCTIONAL PROGRAMMING
Ellis, D.	MEDICAL COMPUTING AND APPLICATIONS
Ennals, J.R.	ARTIFICIAL INTELLIGENCE
Ennals, J.R.	BEGINNING MICRO-PROLOG
Ennals, J.R., *et al.*	INFORMATION TECHNOLOGY AND EDUCATION
Filipič, B.	PROLOG USER'S HANDBOOK
Ford, N.	COMPUTER PROGRAMMING LANGUAGES
Ford, N.J., Ford, J.M., Holman, D.F. & Woodroffe, M.R.	COMPUTERS AND COMPUTER APPLICATIONS: An Introduction for the 1990s
Grill, E.	RELATIONAL DATABASES
Grune, D. & Jacobs, C.J.H.	PARSING TECHNIQUES: A Practical Guide
Guariso, G. & Werthner, H.	ENVIRONMENTAL DECISION SUPPORT SYSTEMS
Harland, D.M.	CONCURRENCY AND PROGRAMMING LANGUAGES
Harland, D.M.	POLYMORPHIC PROGRAMMING LANGUAGES
Harland, D.M.	REKURSIV
Harris, D.J.	DEVELOPING DEDICATED DBASE SYSTEMS
Henshall, J. & Shaw, S.	OSI EXPLAINED, 2nd Edition
Hepburn, P.H.	FURTHER PROGRAMMING IN PROLOG
Hepburn, P.H.	PROGRAMMING IN MICRO-PROLOG MADE SIMPLE
Hill, I.D. & Meek, B.L.	PROGRAMMING LANGUAGE STANDARDISATION
Hirschheim, R., Smithson, S. & Whitehouse, D.	MICROCOMPUTERS AND THE HUMANITIES: Survey and Recommendations
Hutchins, W.J.	MACHINE TRANSLATION
Hutchison, D.	FUNDAMENTALS OF COMPUTER LOGIC
Hutchison, D. & Silvester, P.	COMPUTER LOGIC
Koopman, P.	STACK COMPUTERS
Koops, A. & J. Dreijklufft	WORKING WITH COREL DRAW!
Kenning, M.-M. & Kenning, M.J.	COMPUTERS AND LANGUAGE LEARNING: Current Theory and Practice
Koskimies, K. & Paaki, J.	AUTOMATING LANGUAGE IMPLEMENTATION
Koster, C.H.A.	TOP-DOWN PROGRAMMING WITH ELAN
Last, R.	ARTIFICIAL INTELLIGENCE TECHNIQUES IN LANGUAGE LEARNING
Lester, C.	A PRACTICAL APPROACH TO DATA STRUCTURES

Series continued at back of book

COMPUTERS AND GRAPH THEORY
Representing Knowledge for Processing by Computers

RAMACHANDRAN BHARATH Ph.D.
Walker Cisler School of Business
Northern Michigan University, USA

ELLIS HORWOOD
NEW YORK LONDON TORONTO SYDNEY TOKYO SINGAPORE

To
Kausie Lokanathan,
Ramachandran Sivasankar,
Rama Rajagopal,
P. V. Rajagopal.

First published in 1991 by
ELLIS HORWOOD LIMITED
Market Cross House, Cooper Street,
Chichester, West Sussex, PO19 1EB, England
A division of
Simon & Schuster International Group
A Paramount Communications Company

Typeset in Times by Ellis Horwood Limited
Printed and bound in Great Britain
by Redwood Press Limited, Melksham, Wiltshire

British Library Cataloguing in Publication Data

Bharath, Ramachandran
Computers and graph theory: Representing knowledge for processing by computers.
(Ellis Horwood series in computers and their applications)
I. Title II. Series
511.0285
ISBN 0–13–161605–6

Library of Congress Cataloging-in-Publication Data available

Table of contents

Preface

Dr Eugene Wigner† gave the memorable title, 'The unreasonable effectiveness of mathematics in the natural sciences', to a famous article in which he set out his thoughts on the apparent miracle that the abstract concepts of mathematics turn out to have so many concrete applications. The branch of mathematical thought known variously as the theory of graphs or networks seems indeed to be something which is extremely 'unreasonably effective'. It starts with something which appears very simple, the identification of relationships between one or more (possibly all) pairs of objects in some collection of objects, and representing:

(1) the objects as circles
(2) the relationship between pairs of objects by connecting lines.

The classic example, with which this book starts, is of cities and interconnecting roads, where pairs of cities are linked by roads. But abstracting from this, it turns out that the theory captures something fundamental about our knowledge of the world, namely that knowledge often (always?) consists of relationships between objects, and of tracing the implications of such relationships.

The first aim of this book is to give an easy-paced introduction to the fundamental concepts of the theory of graphs. The book does not attempt in any way to be comprehensive — the subject is too vast for that. What it aims to do is to provide a representative sample of the ways in which graph theory

† Wigner, Eugene P.: The unreasonable effectiveness of mathematics in the natural sciences, *Communications in Pure and Applied Mathematics*, **13** (1960): 1–14.

can be applied for representing knowledge and solving problems. I hope that the sample is sufficient to provide the basis for further study in the subject.

The second aim of the book is to show how the theory of graphs lends itself very well to computer implementation. The declarative language Prolog is one which is very suitable for expressing the problem-solving methods which emerge from the theory of graphs. One advantage of Prolog is that mastery of a small subset of the language enables one to tackle and solve a variety of problems, after a short initial learning period. I hope the sample programs given will enable the user to get a working knowledge of Prolog, but the emphasis is on Prolog as a vehicle for implementing the ideas discussed.

I have given a short guide to further reading in the last chapter. I hope the book will be of use as supplementary or recommended reading in courses in data structures, programming languages and logic programming. I further hope that I have achieved my aim of writing the book in such a way that persons studying on their own would be able to use it.

ACKNOWLEDGEMENTS

I wish to thank the following:

— TAB BOOKS Inc. Blue Ridge Summit, PA, U.S.A. for permission to use some of the material from my book *Prolog* (TAB BOOK Number 3902) published by them in 1989.

— Ms. Donna Yelle for the fine job she did of preparing the various diagrams for the book.

1

Transition networks

PREVIEW

In this chapter, we will look at two important ideas which are used throughout the rest of the book:

(1) the technique of representing knowledge about relationships between various objects in the form of a **transition network**.
(2) the concept of **recursive definitions** or programs, with special reference to the problem of tracing paths in a network.

TRANSITION NETWORKS: AN EXAMPLE

Let us consider a network of roads connecting a number of cities, called city_1, city_2, city_3 . . . , with direct links between some (maybe all) pairs of cities. We can represent the network as in (Fig. 1.1) with the cities shown as circles and the roads as lines connecting the circles. The roads have labels which show the distances. To start with, we think of the roads as one-way, in the direction shown by the arrows. We can easily extend the model to include two-way roads as well — for example, if city_1 to city_2 is a two-way road, we can draw another link between city_2 and city_1, in addition to the link between city_1 and city_2. I have labelled the cities as city_1 etc., using lower case rather than upper case for the first letter, and connecting the word city with a number using an underline character rather than leaving a blank. This is to make it more convenient to develop Prolog programs later in the chapter. In Prolog, upper case letters have a special significance, as we will see later, and blanks are not allowed in the names of specific objects.

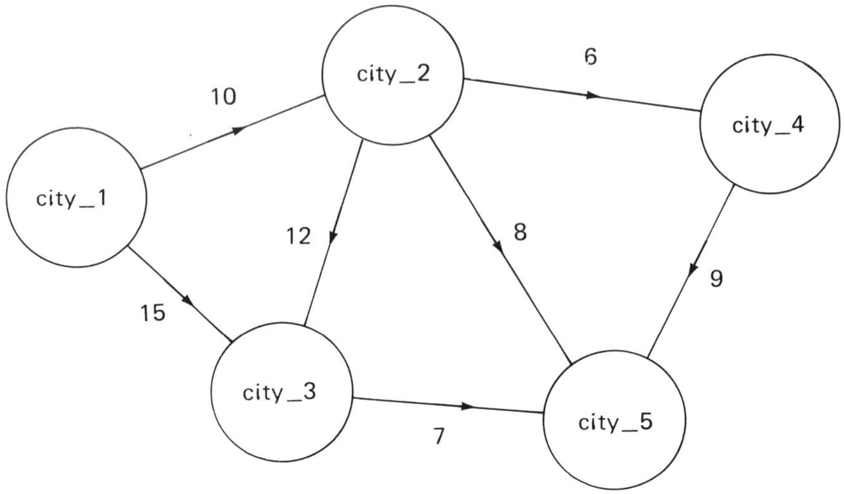

Fig. 1.1.

Suppose we want to find all possible routes between cities which are not directly linked, say city_1 and city_5. One way of describing and thinking about the problem is as follows:

(1) to begin with, we have a **starting state** or **initial state** city_1, and we are aiming for a **goal state** city_5.
(2) From city_1, it is possible to change to a new state city_2 or city_3. From city_2, it is possible to go to state city_3, or city_4 or city_5. Similarly, from any particular state, there are certain changes of state possible.
(3) We want to find all the alternative routes from city_1 to city_5. In terms of changes of state, we can describe each path as a series or sequence of changes of state, leading from the starting state to the goal state. An extension of the problem would be to compare the different paths and find the shortest one.

This way of describing and analysing a problem in terms of various states and changes of state is called the state space model, because it deals with a collection or **space** of different states. This conceptual model is also often called a **finite state machine** in books on automata theory, languages and grammars. The word machine is used in an abstract sense, in roughly the same sense as 'system'. The state space model or finite state machine model turns out to be a handy conceptual tool which can be applied in a variety of different situations, as we will see from other examples we will look at below.

Further, it turns out to be a convenient way for representing such problems for processing by computers, and this book is essentially about showing how this way of representing knowledge is very useful for use in computer implementations of problem solving.

The general name for a circle representing a state is a **node**, and a line connecting two nodes is called an **arc**. The diagram showing nodes and arcs is called a **transition network**, since it shows the possible transitions between pairs of nodes. Another commonly used name for this type of diagram is **directed graph**.

A SECOND EXAMPLE: SENTENCES

We can use a transition network for a very different kind of problem — analysing the structure or pattern of sentences in English or other languages. Let us consider the type of sentences which consist of a noun phrase followed by a verb phrase. We can think of a sentence of this kind as being built up in stages, starting in the state of being an empty sentence. A transition network for this would be as in Fig. 1.2.

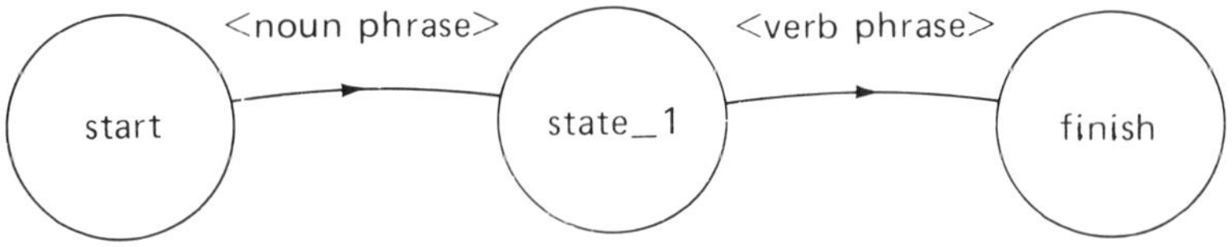

Fig. 1.2.

In the previous example, the label on each arc was a distance, and we can think of it as describing an action, namely travelling 10 miles, which is associated with the change of state from city_1 to city_2. In this example, we can similarly think of the label as associated with a change of state — the first stage in building up the sentence can be thought of as first putting in a noun phrase, and then adding a verb phrase for the next change of state. However, there is one difference. In the previous example, the distances like 10 miles were simple objects, without any further substructure. But in the current example, a term like ⟨noun phrase⟩ has been enclosed in angle brackets to show that it is not a simple object, but in turn has a structure which can be shown by a different transition network. The network for **noun phrase** would be as in Fig. 1.3. Article, adjective and noun would have their own networks. The network for **article** would be as in Fig. 1.4. The network for noun phrase

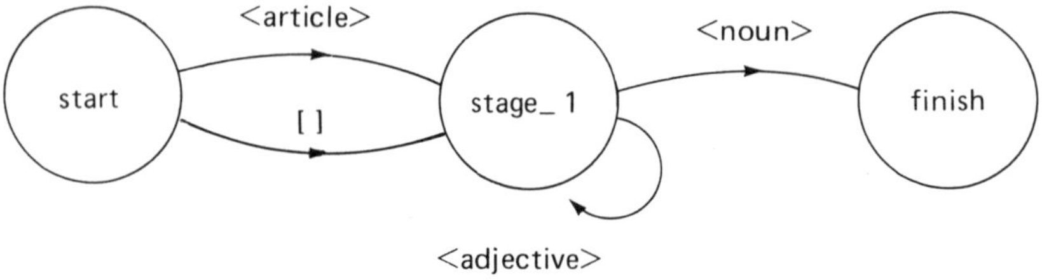

Fig. 1.3.

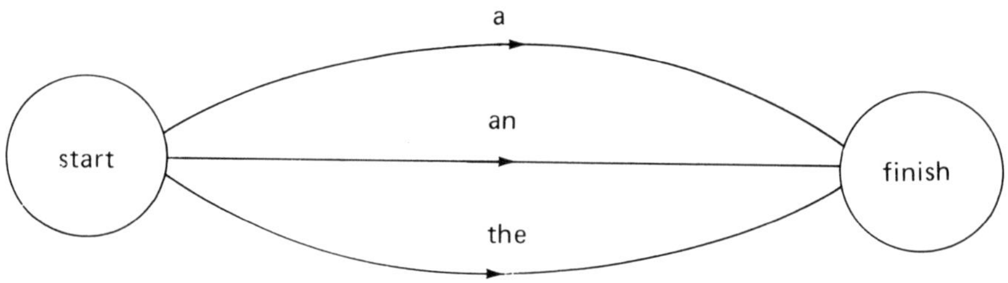

Fig. 1.4.

represents the idea that a noun phrase may start without an article, or with one of the articles, a, an or the, and may have zero or more adjectives, followed by a noun.

The two examples above are from very different areas or domains, but the general model of a transition network fits them both. Let us look at one more example from an entirely different domain. We will look at various other examples in later chapters of the book.

ANOTHER EXAMPLE: A PUZZLE

There is a well-known puzzle about three jugs, jug_a, jug_b, and jug_c, with capacities of 8, 5 and 3 litres respectively. Jug_a is full and the others are empty. No other measuring instrument is available. The problem is to find a series of transfers of liquid between pairs of the jugs so as to end up with 4 litres each in jug_a and jug_b. To apply the state space model to this problem, we can think of the initial state as a list of three numbers, [8,0,0], showing the quantities of liquid in each of the three jugs. The goal state would be [4,4,0]. From the initial state, one can pour from jug_a into jug_b till the latter is full, leading to the state [3,5,0]. Alternatively, one could fill upjug_c, leading to the state [5,0,3]. We can represent this by a transition network as in Fig. 1.5.

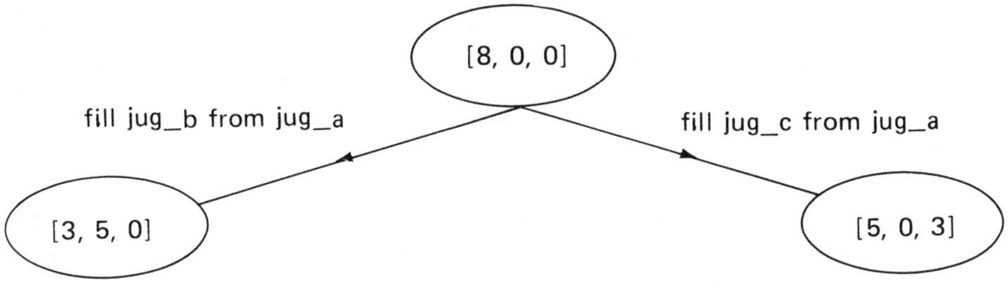

Fig. 1.5.

Going further, from the state [3,5,0] we have possible transitions to states [0,5,3] or [3,2,3] as in Fig. 1.6.

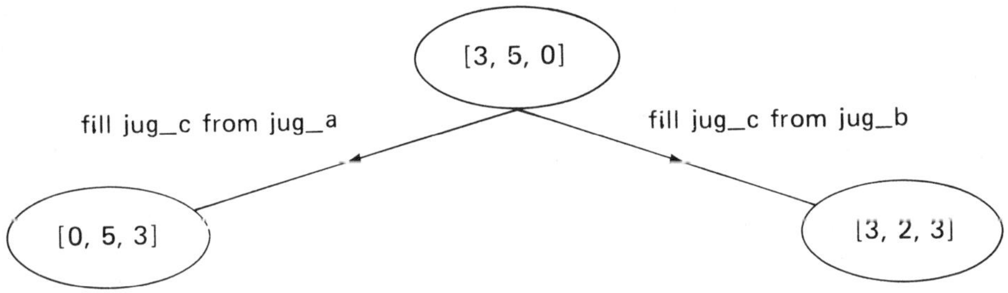

Fig. 1.6.

The solution to the problem would be finding a path which leads from [8,0,0] to [4,4,0].

PATHS AND RECURSION

In the three examples we have seen above, solving the problem required tracing a path in the network, whether the path was made up of roads, categories of words or actions representing pouring liquid from one jug to another. This task of finding a path comes up in all problems which can be modelled as transition networks.

Let us look at the task of finding a path from a given start node, city_1, to a finish node city_finish, in a network of cities and roads. We will think of the network of five cities which we saw in our first example as extended to include many more cities and roads. One way of describing the path from city_1 to city_finish is to say that the path consists of an arc from city_1 to a directly connected node, say city_2, followed by an arc from city_2 to a directly connected node, say city_3, and so on, until the goal city_finish is reached. This description suggests that the technique for solving the problem would be to repeatedly 'find the next node', first finding the node next to city_1, then the node next to that node . . . until city_finish is reached.

A different way of describing the path from city_1 to city_finish is to say that it consists of an arc from city_1 to a next node, followed by a path from that next node to city_finish. The interesting feature of this alternative description is that in the description or definition of the relationship 'path' between city_1 and city_finish, we have used the same concept of 'path', but between a different pair of objects city_2 and city_finish. Such a description is called **recursive**, which means 'returning to itself'. A recursive specification of a problem suggests that the solution procedure would be to identify subproblems of the same type as part of the original problem, and use the same solution procedure for the subproblems. That is, to find a path from city_1 to city_finish, we find a path from city_2 to city_finish and use that as part of the solution for the original problem. Or more generally, to solve a problem, we identify a subproblem1 whose solution is part of the solution of the problem we start with. This means in turn that we have to find a subproblem2 of the same type as part of subproblem1, and again a subproblem3 as part of subproblem2 . . . Ultimately, the subproblem of the subproblem of the subproblem . . . of the original problem becomes a very simple problem for which the solution is obvious or straightforward. In the example we are considering, finding a path from city_1 to city_finish requires us to solve the subproblem of finding a path from city_2 to city_finish, which in turn leads to the subproblem of finding a path from city_3 to city_finish . . . Now suppose after a number of steps we come to the subproblem of finding a path from city_p to city_finish, which requires finding a path from city_q to city_finish, which requires finding a path from city_finish to city_finish as shown in Fig. 1.7. Finding a path from city_finish to city_finish is a subproblem with an obvious solution: just stay at city_finish. This simple problem which is the end point of reducing a problem recursively to subproblems is called the **non-recursive case** or the **boundary case**. Having reached this, the solution to the original problem is found by 'unwinding the recursion'. In this case, since we have the answer for the path from city_finish to city_finish, we use this to find the answer for the path from city_q to city_finish, and this answer in turn completes our answer for the path from city_p to city_finish . . . until we have the complete answer for the path from city_1 to city_finish. One can think of recursion as a strategy in which the solution for the

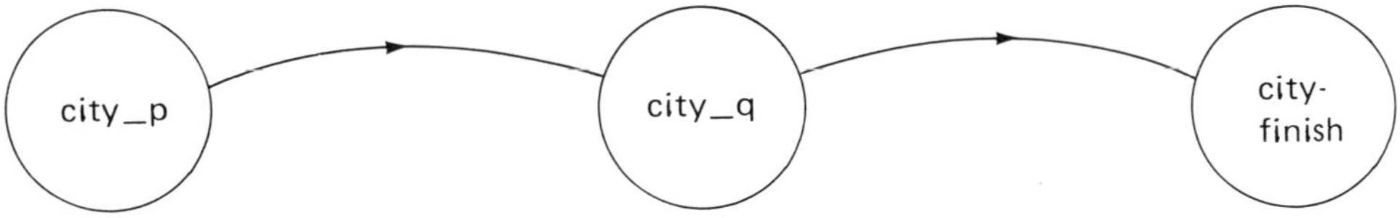

Fig. 1.7.

simplest, or boundary case, is used as a building block for the solution for the next simplest case, a path with one arc, and this in turn is used for the solution for the next simplest case, a path with two arcs, and so on.

The advantage of specifying the solution for problems in a recursive way is that it makes it possible to write very concise and clear programs in programming languages that support recursion. As an example, let us look at Prolog, which is very convenient for recursive solutions in general and in particular for problems involving networks.

PROLOG AND A PROGRAM FOR TRACING PATHS IN NETWORKS

Let us now look at how networks are represented in Prolog, and how the recursive definition of a path can be programmed in that language.

As is true of any programming language, Prolog has its own rules about how to express or represent the relevant knowledge for solving a problem so that it can be communicated to a computer which has been equipped to handle that language. Prolog was designed to make the form in which we communicate knowledge to a computer as close as possible to the description we would give in English. In this case, one of the things we want to represent is the fact that there are pairs of nodes city_1 and city_2, city_1 and city_3, and so on, which are related in a way we describe by saying there is an arc between them. We also want to express the fact that each arc has a certain distance associated with it. We would express this in a Prolog program in the following form:

```
arc(city_1,city_2,10).
arc(city_1,city_3,15).
arc(city_2,city_5,8).
arc(city_2,city_4,6).
arc(city_2,city_3,12).
arc(city_3,city_5,7).
arc(city_4,city_5,9).
```

and so on, if there are more nodes and arcs.

Each of these statements is called a **clause** and the collection of various clauses which we provide the system with is called a **knowledge base**. The clauses above convey information about a relationship 'arc', between objects city_1 and city_2, city_1 and city_3, and so on. In Prolog, the label or name of the relationship is called the **predicate** of the clause, and the objects or entities which are involved in the relationship are called the **arguments** of the clause. Each clause ends with a period.

When a program for handling Prolog has been loaded into a computer, we can think of the computer plus this program as a 'Prolog system', or alternatively, we can say the computer is now functioning in the 'Prolog mode', or as a 'Prolog machine'. When it is doing so, and we give it information of the type given above, the computer can use this information to answer various types of questions from the user:

(1) Firstly, it can verify and confirm that something the user is asking for is in fact contained in the program's knowledge base. For instance, if the user types in:

    ?-arc(city_3,city_5,7).

    the system will check if this matches any clause it holds in the knowledge base, and will respond 'yes'. Similarly, the response to a question:

    ?-arc(city_2,city_3,18).

    will be 'no', since the question does not match the clause arc(city_2, city_3,12). which is in the knowledge base. Just as in the case of clauses in the knowledge base, the questions put by the user have to end with a period. Most Prolog systems would show the ?- as a prompt on the screen, for the user to type in questions.

(2) The system can do something more than just checking whether or not the question matches something contained in the knowledge base. It can respond to questions asking it to extract information from the knowledge base, if it is given some clues as to what to search for. For example, if the user asks:

    ?- arc(city_2,city_3,What).

    the system would respond What=12. The point to note is that because the user has used 'What' *starting with an upper case letter W*, the system takes this to be a **variable**, which takes on appropriate values as necessary. In other words, the system takes the user's question to mean:

What is the distance between city_2 and city_3? Clearly, the knowledge base provides the necessary information that this distance is 12, and so the variable What is given a value of 12. The user could have asked:

```
?-arc(city_2,city_3,Q).
```

and got the response Q=12. Similarly, the user could use any word starting with an upper case letter instead of What, but one would naturally like to use a word like What or Distance, rather than Q or X, to make the response immediately understandable. As another example, if the user asks:

```
?-arc(city_1,Endpoint,Distance).
```

the system takes this to mean that the user want to know the endpoint of an arc starting at city_1, and the distance. In the sample knowledge base above, this question has more than one answer, because it can match more than one clause, and the system would give all the possible answers in succession:

```
Endpoint=city_2    Distance=10
Endpoint=city_3    Distance=15
```

It is important to keep in mind that the system always starts its search at the top of the knowledge base and goes down the knowledge base, following the order in which we have put in the clauses. Usually, most Prolog systems are designed to show the first possible answer on the screen, and then wait for the user to indicate by typing in a ; that the second possible answer should now be shown . . . and so on.

(3) When the user asks questions like ?-arc(city_1,city_2,15)., or ?-arc (city_1,city_3,Howfar)., it is usual to describe this by saying that the user is setting a **goal** for the system to **satisfy**. It may be able to satisfy the goal and give an answer, either of 'yes', or of values for variables which have been matched with specific values. Or, it may not be able to satisfy the goal, in which case the system gives the answer 'no'.

Program for tracing paths

We want to tell the computer that we would like to trace paths from any given starting node, which we may call Start, to any end node, to be called End.

The recursive strategy we have seen above is to first find an arc from Start to any other node, which we may call Next, and then find a path from Next to End. We use names beginning with upper case letters, since we want to use the program for any starting node and any ending node.

To express this in Prolog, we would put in the following clauses in the knowledge base:

```
/* NON-RECURSIVE CASE or BOUNDARY CASE */
   route(Same,Same,[Same]).

/* RECURSIVE CASE */
   route(Start,End,[Start|Rest]):-
            arc(Start,Next,Distance),
              route(Next,End,Rest).
```

Any statement put into a knowledge base between the symbols /* and */ is treated by the Prolog system as something which is inserted as a **comment** for the human reader, and not a statement to be acted on by the system. Some Prolog systems have a different convention, using a % symbol instead of the above symbols for enclosing comments.

Let us first look at the statement or clause which follows the comment **recursive case**. To see what it means, we need to first look at some of the conventions for representing knowledge in **Prolog**.

(1) One convention is that a list of items is represented by writing down the items separated by commas, and enclosing the whole list in rectangular brackets. For example, a list of three names would be represented as:

    [mary,john,elizabeth].

    The items in the list are called the elements of the list.

(2) A second feature of lists in Prolog is that a list can contain another list as an element, e.g. the list:

    [[a,b],c,[d,e,f,g]]

    is a list with three elements in it, in which the first and third elements of the list are themselves lists, containing respectively two and four elements, whereas the second element in the list is just the element c.

(3) A third point to note about lists in Prolog is that we distinguish between the first element of the list, called the **head** of the list, and the rest of the list, called the **tail** of the list. In the list [a,b,c,d], a is the head and [b,c,d] is the tail.

This distinction is shown by using a symbol | to separate the head and the tail of a list when writing Prolog programs. Also, there may be situations where we start to build up a list from scratch, or remove elements from a list till we are left with nothing. So, we need something to represent an empty list, and Prolog uses [] for this purpose.

(4) Another point to note is that the symbol :- signifies that the statement is a **conditional clause**, in which the truth of what is first stated depends on the conditions which follow the symbol. We could read a conditional statement as saying 'such-and-such is true WHERE the following conditions are true'. In contrast to this conditional clause, the earlier clauses we looked at, such as arc(city_1,city_2,10) were **unconditional clauses**. Sometimes unconditional clauses are referred to as **facts** and conditional clauses as **rules**.

(5) We looked earlier at the concept of the user setting goals for the system to satisfy. In the case of conditional clauses, we can describe them as having one goal in the **head of the rule**, the part before the :- symbol, and also a set of goals in the **body of the rule**, or the part following the :- symbol. If the goal set by the user has the same predicate as the **head goal** of a conditional clause, the system tries to satisfy all the goals in the body of the clause, proceeding from left to right.

Now, let us look at the **head** of the conditional clause or rule for the recursive case:

```
route(Start,End,[Start|Rest]) :-
```

In the light of the features of Prolog which we have discussed above, we can see that it says that the route from a node Start to a node End is a list with Start at the head, and Rest as the tail, where the value of Rest would be given by the **body** of the rule :

```
:-
    arc(Start,Next,Distance),
      route(Next,End,Rest).
```

The body of this rule has two goals in it, separated by a comma, and as mentioned earlier, the system will try to satisfy these goals one by one.

Looking at these two goals, we can see that that they say that the value of Rest is derived as the route from Next to End, where Next is a node such that there is as an arc from Start to Next. In other words, it is more or less a direct translation of the way we specified the problem in English. I have changed the name of the relationship, or predicate, to route instead of path to make it specifically suitable for the problem of a route between cities which we are looking at. Let us look at what happens if we ask:

```
?-route(city_1,city_5,What).
```

(a) the Prolog system will try first the non-recursive clause, but will find that it does not fit. The reason is that this clause requires that both the objects (or in this case, nodes) given in parentheses after the word 'route' should be the same, so that they can be represented by a single variable Same. But in our question, city_1 and city_5 are different objects.

(b) The system will therefore now try the recursive rule. First, it will match the user's question and the head of the rule:

```
?-route(city_1,city_5,What).
  route(Start,End,[Start|Rest]).
```

So it will identify the variable Start as standing for city_1, the variable End as standing for city_5, and the variable What as standing for the list [city_1|Rest], i.e. a list with city_1 as the head, and a tail which is a variable Rest. In Prolog terminology, the variables Start and Finish are said to be **instantiated** or **bound** to the values city_1 and city_5 respectively. The variable What is not yet bound, because it is matched with a list in which a part, i.e. Rest, is itself a variable. As soon as Rest becomes bound, What will also become bound. So, the system cannot give us an answer at this stage, and in any case, it will not give back an answer until it has processed all the goals in the body of the clause,which follows the WHERE symbol, :-.

The body of the clause has in this case two separate goals, separated by a comma. The first goal in the body is now in effect arc(city_1,Next,Distance), because the variable Start has already been instantiated. The system searches the sample knowledge base above, *following the order in which we have entered the clauses.* We see that the first match it finds in its search will be a match between the following:

```
arc(city_1,Next,Distance).
```

```
arc(city_1,city_2,10).
```

As a result, Next is bound to the value city_2. The next goal in the body of the clause tells the system to try the goal route(city_2, city_5,Rest). Once again, it first tries to see if it can satisfy this goal with the first clause with predicate 'route' which it can find in the knowledge base, i.e. the non-recursive clause. Again, it cannot use this clause, because city_2 and city_5 are different. So, it tries again with the recursive clause.

(c) It is helpful in understanding how Prolog works to think of the system as making a new 'copy' of a recursive rule whenever it needs to use the same rule again, and building up a 'stack' of such copies in the memory of the computer. So, we can imagine this second copy as being of the form:

```
route(Start2,End2,[Start2|Rest2]):-
          arc(Start2,Next2,Distance2),
            route(Next2,End2,Rest2).
```

Since this is a second copy of the rule, we may think of it as having the variables now modified with 2 added at the end. We can think of this as put at the second level of the stack:

....
:

Level 3 of the stack

Level 2 of the stack

Level 1 of the stack

By the same line of reasoning as before, we can think of Start2 as bound to the value city_2, and End2 to the value city_5. The variable Rest is matched with [city_2|Rest2], but is not yet bound because Rest2 is yet to be bound.Again, we can see that as it searches down the knowledge base, the system will bind Next2 to city_5, and try to satisfy a new goal:

```
route(city_5,city_5,Rest2).
```

At this stage the non-recursive clause will be triggered, because there is a match possible between the following:

route(city_5,city_5,What).
route(Same,Same,[Same]).

So, the system finds [city_5] as the value to which Rest2 is bound.

This search for the value of Rest2 started at level 2 of the stack. The goal in the body of the rule at this level of the stack was route(city_5,city_5,Rest2) and the goal in the head of the rule was route(city_2,city_5,[city2|Rest2]). So the system now has an answer for [city_2|Rest2] as the list:

[city_2|[city_5]],

which is the same as

[city_2,city_5]

The task at level 2 is now complete, so the system can 'unstack' or remove the copy of the rule at this level, and pass back the answer it has found to level 1.

The recursive call to level 2 arose because, at level 1, the head goal had matched the variable What in the user's request with [city_1|Rest], and was trying to find the value of the variable Rest from the body of the rule at level 1, by calling the goal:

route(city_2,city_5,Rest).

where it was trying to find an answer for the variable Rest from level 2 of the stack, by matching it with [city_2|Rest2]. The system can therefore complete the head goal at level 1 of the stack by binding the variable What to the list:

[city_1|[city_2,city_5]]

which is the same as

[city_1,city_2,city_5]

The system now gives the answer:

What=[city_1,city_2,city_5].

Now, what happens if we ask the system to find any alternative route which may be there ? The last step in the previous solution was that the system found an answer to route(city_5,city_5,Rest2). using the non-recursive clause. So, if we ask for another solution, it will try this goal again on the recursive clause this time, and obviously will fail. So, it will **backtrack** to its previous matching of arc(city_2,Next2,Distance2). with arc(city_2,city_5,8). and go down the knowledge base to find another match. We can see that it will find this new match with arc(city_2,city_4,6). As a result it will now try the new goal arc(city_4,city_5,Rest2). This time we see it will again have to use a third copy of the recursive rule:

```
route(Start3,End3,[Start3|Rest3]):-
                arc(Start3,Next3,Distance3),
                  route(Next3,End3,Rest3).
```

where Start3 and End3 will be bound to city_4 and city_5, and Rest2 will be matched with [city_4|Rest3], and Next3 will be bound to city_5, and a new goal route(city_5,city_5,Rest3) will be tried. This goal can be satisfied using the non-recursive rule, so that Rest3 becomes [city_5], and consequently Rest2 becomes [city_4,city_5], Rest becomes [city_2,city_4,city_5] and What becomes [city_1,city_2,city_4,city_5]. If we ask for more solutions, the system will find the other routes

[city_1,city_2,city_3,city_5]

and

[city_1,city_3,city_5]

as well.

If we ask for a route which involves many more arcs, the program would make many more copies of the clause, one for each recursion, and finally find the solution.

The advantage of a language like Prolog which comes with convenient features for handling lists is that we can easily extend the programs using more lists. Suppose we want our 'route' program to build up a list of the distances of the various arcs in the routes it finds. All we need to do is to put in a fourth argument which will build up a list of the lengths of the arcs, just as the third argument builds up a list of the nodes. The modified program would be:

```
/* BOUNDARY CASE */
route(Same,Same,[Same],[]).
/* RECURSIVE CASE */
route(Start,Finish,[Start|Rest],[Distance|Distrest]):-
     arc(Start,Next,Distance),
        route(Next,Finish,Rest,Distrest).
```

If we now ask:

```
?-route(city_1,city_5,What,Howfar).
```

we would get back the answer:

```
What=[city_1,city_2,city_5]   Howfar=[10,8].
```

We will get back later to the question of how to add up the various distances so as to get the total distance.

To conclude, let us consider the following problem. The arcs we have put into the knowledge base are all one-way. Now suppose we put in two-way arcs, so that the knowledge base has arc(city_1,city_2,10). and also arc(city_2,city_1,10)., and let us put these two as the first two clauses in the knowledge base. If we now ask: ?-route(city_1,city_5,What)., as before, the program uses arc(city_1,city_2,10). and recurses to route(city_2,city_5,Rest). When it tries this goal, it will use arc(city_2,city_1,10). because this is the very first one it can find in the knowledge base to match arc(city_2,Next2,Distance2). As a result, it will recurse to route(city_1,city_5,Rest2). We can see that the program will never get past the two clauses, and will go on for ever — or rather, we will get a message on the screen which says 'out of stack space'. We saw earlier that a recursive program can be thought of as making copies of the recursive clause, one for each use of the clause, and that these copies can be thought of kept in a stack, one on top of the other, in the memory of the computer. When the answer to the topmost copy has been found, this copy can be taken off the stack, and the answer used in the previous copy which is now at the top of the stack. Here, because the recursion goes on and on, ultimately there is no space left in the memory of the computer for stacking more copies of the recursive rule, and the system tells us so.

To prevent this kind of thing happening, we need to check that in finding a route, the program is not going back to one of the nodes already included in the route. One way to do this is to use one more argument in the program, which will contain all the nodes found so far, and each time a new 'Next' node is found, the program will check to see that this is not a node already visited. A modified program for this is given below:

```
/* THIS PROGRAM CALLS AN AUXILIARY PROGRAM */
route2(Start,Finish,Answer) :-
            routea(Start,Finish,Answer,[Start]).


/* AUXILIARY PROGRAM */

/* BOUNDARY CASE FOR routea */
routea(Same,Same,Answer,Answer).

/* RECURSIVE CLAUSE FOR routea */
routea(Start,Finish,Answer,Sofar):-
        arc(Start,Next,Distance),
          not (member(Next,Sofar) ),
      routea(Next,Finish,Answer,[Next|Sofar]).
```

When the user asks the question ?-route2(city_1,city_5,What)., we see that this leads to the auxiliary program 'routea' starting off with ?-routea-(city_1,city_5,Answer,[city_1]). Now, the non-recursive clause for routea will not match the question, so the recursive clause is tried.

The next step is the same as in the previous analysis, with Next being bound to city_2. But now there is one more goal in the body of the clause, before the recursion starts. Most versions of Prolog have a built-in predicate 'member', which takes two arguments, the second of which has to be a list. As the name suggests, the predicate checks whether the first argument is one of the elements or members of the second argument. In this program, we have added a 'not'. Since 'Sofar' has been bound to [city_1], in effect the program checks to see if it is true that:

```
not(member(city_2,[city_1])).
```

This is true, because city_2 is not in the list [city_1], so the program proceeds further, and recurses to

```
routea(city_2,city_5,Answer,[city_2,city_1]).
```

This time too the boundary case does not apply, so the program searches the knowledge base and finds arc(city_2,city_1,10). and binds Next2 (this is the second copy of the clause which the program is now using) to city_1. But, when it tries:

```
not(member(city_1,[city_2,city_1])
```

this goal fails, since city_1 is in the list [city_2,city_1].

So, the program backtracks and searches further and finds another match for arc(city_2,Next2,Distance2), i.e. arc(city_2,city_5,8). This time, since city_5 is not a member of [city_2,city_1], the program can proceed further and recurse to routea(city_5,city_5,Answer,[city_5,city_2,city_1]).This goal can be satisfied using the boundary case clause, so Answer gets bound to [city_5,city_2,city_1]. Since What in the user's question was matched with Answer, the system responds with

```
What=[city_5,city_2,city_1].
```

The list is the reverse of the route from city_1 to city_5, but this is no problem, because many versions of Prolog have a built-in predicate 'reverse' and we could use this in the boundary case:

```
routea(Same,Same,Answer,Sofar) :-
        reverse(Sofar,Answer).
```

Both the member program and the reverse program are based on interesting principles which illustrate various features of Prolog. So, I have given an appendix showing how one can build the member program from first principles if one has a version of Prolog which does not provide it as a built-in predicate.

The techniques we have seen above, of using an auxiliary program with an additional argument (Sofar) for gathering intermediate results as the recursion goes on are of general use in dealing with network problems, and we will use these in various other places in the book.

SUMMARY

This chapter has

(1) introduced the key concept of transition networks as a technique for representing knowledge
(2) emphasized the importance of recursion as a way of looking at the task of tracing routes in networks
(3) indicated how Prolog is a handy tool for representing networks and tracing routes.

In subsequent chapters, we will look at the applications of these concepts in a variety of areas. For the benefit of readers who are encountering Prolog for the first time, I give below a short list of questions for them to think about. Suggested answers are at the back of the book.

QUESTIONS

(1) What difference, if any, would it make if we changed the non-recursive clause for route to read:

  (a) route(End,End,[End]).
  (b) route(Start,Start,[Start]).

(2) When we frame a recursive definition in situations involving lists, we try to make the recursive use of the definition apply to the tail of the list, thence to the tail of the tail of the list . . . till we end up with a boundary case.

  How can we apply this approach to define the last element of a list, and how can we code this in Prolog?

APPENDIX

Let us first develop a recursive definition of what it means to say that a particular element is a member of a list, and then put it in Prolog.

We can say that an element is a member of the list, if

(a) it is the head of the list, or

(b) it is a member of the tail of the list.

(a) can be expressed in Prolog as:

```
member(Element,[Element|_]).
```

Here the new feature is that we use an underline character for the tail of the list, instead of a specific variable name. This is because we are not interested in the value of the tail of the list — if the element we are looking for matches the head of the list, then what is in the tail is irrelevant. The underline character used this way is called an **anonymous variable**.

(b) can be expressed as:

```
member(Element,[_|Tail]) :-
    member(Element,Tail).
```

This is the recursive part of the definition. This second clause will be tried by the system only if the first clause does not fit, i.e. if the element we are looking for does not match the head of the list, so now we are not interested in the head of the list, and make it an anonymous variable.

?- member(a,[a,b,c]). would immediately be answered 'yes' using clause (a). ?-member(d,[a,b,c,d]). would be answered 'yes', after recursions through member(d,[b,c,d]). member(d,c,d]). and member(d,[d])., using clause (b).

This program also illustrates one of the nice features of Prolog, its flexibility. If we ask ?-member(What,[a,b,c]). all the possible answers, What=a, What=b, and What=c, will be produced in succession, which is what we would intuitively expect as well.

As another illustration of the idea of recursion, let us get back to a question left unanswered earlier. The program for finding paths gave us the lists of nodes and distances in the path, as in the example considered in the body of the chapter, where we would invoke the program with a question like:

```
?-route(city_1,city_5,What,Howfar).
```

and get a response:

What=[city_1,city_2,city_5] and Howfar=[10,8].

Let us consider the recursive definition of a predicate which, given a list of numbers, would give us the sum, so that we could use it with a list like [10,8] to get the sum 18.

We can recursively define the sum of the numbers in the list as equal to the head of the list plus the sum of the numbers in the tail of the list. We can see that this will recurse to the state where the list becomes empty, and in that case the sum is of course 0. Based on the principles we have seen so far, we can define a predicate 'addup' for this as follows:

```
/* BOUNDARY CASE */
    addup([],0).
/* RECURSIVE CASE */
addup([H|T],Answer) :-
    addup(T,Tailsum),
    Answer is H+Tailsum.
```

We can use this module in conjunction with our previous program, which had the predicate route, and which we invoked by asking: route(city_1, city_5,What,Howfar)., by introducing an auxiliary predicate 'routedist':

```
routedist(Start,Finish,Route,Distance) :-
    route(Start,Finish,Route,Distlist),
        addup(Distlist,Distance).
```

We can see that now asking:

```
?-routedist(city_1,city_5,What,Howfar).
```

should lead to the response:

```
What=[city_1,city_2,city_5] and Howfar=18.
```

2

Solving puzzles

PREVIEW

Solving puzzles is traditionally regarded as an exercise of intelligence. In this chapter, we will see how the concept of transition networks discussed in the first chapter can be used for solving two well-known puzzles. We will also develop Prolog programs for those solutions.

THE BRIDGES OF KONIGSBERG

In fact, the use of networks consisting of nodes and arcs originated in solving a puzzle and is generally attributed to the eighteenth-century Swiss mathematician Leonhard Euler (1707–1783), who used it to solve the famous **Seven bridges of Konigsberg** puzzle.

In the Prussian city of Konigsberg (now known as Kaliningrad), the river Pregel flowed through the town, enclosing an island called the Kniephof. The island was shown as A in Euler's diagram in his paper in which he addressed the problem. The island was connected by two bridges to each of the banks (C and D in the Fig. 2.1) of the river, and in addition to these four bridges, there were three more as shown in Fig. 2.1, connecting the banks and the island to another area B. The puzzle which the citizens of Konigsberg tried to solve was this: was it possible to take a walk in such a way that one would cross all of the seven bridges using each bridge exactly once, and come back to one's starting point? Nobody was able to find a way to do it, but that did not of course prove that it was impossible.

Euler solved the problem essentially by representing the problem in terms of a network, in which there were four nodes for the two banks and for

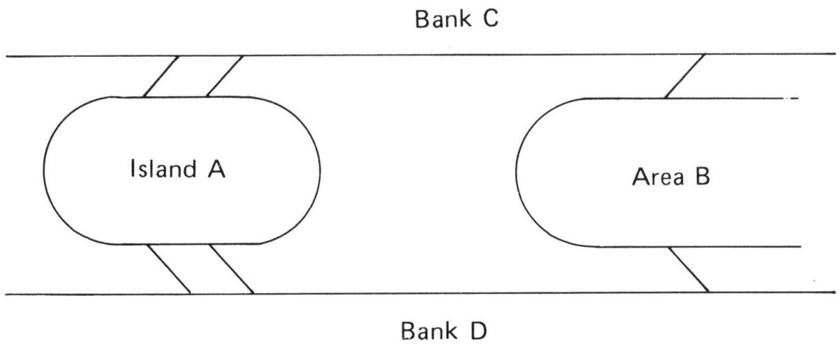

Fig. 2.1.

the island and the area B which the bridges led to, and seven arcs for the bridges, as in Fig. 2.2.

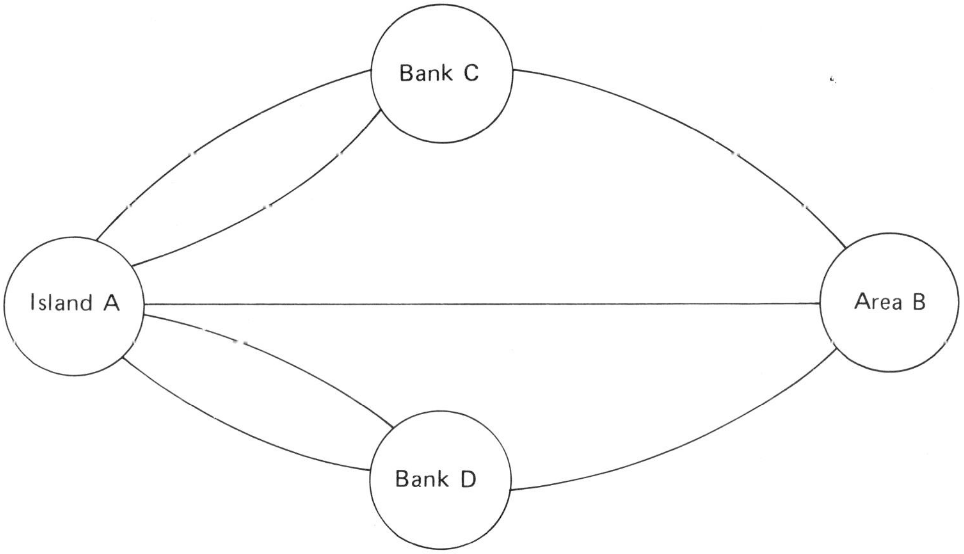

Fig. 2.2.

Euler generalized the problem to consider any finite number of nodes, instead of just four, and any number of interconnecting bridges, instead of seven, and considered the requirements for being able to take a walk which covered each bridge exactly once. Firstly, he pointed out that for such a walk to be possible, each of the nodes, including the starting node, must have an even number of bridges connected to it. This is necessary, because if one uses bridge 1 to first leave the starting point, bridge 2 to come back again, bridge 3 to leave the starting point again and so on, if there are an odd number of bridges, it is clearly not possible to come back to the starting point after using each bridge exactly once. The same argument applied to every node leads to the conclusion that unless every node has an even number of bridges connected to it, a walk of the type required is not possible. So, as far as the particular example of the bridges of Konigsberg is concerned, the answer is no, one cannot take a walk of the kind required.

Euler went further and showed that the condition about every node having an even number of bridges is not merely necessary, it is sufficient as well, i.e. if there are an even number of bridges for each node, it is definitely possible to start and end at the same node after using each bridge exactly once. He further proved that if two of the nodes, say A and B, have an odd number of bridges connected to them, and all the others have an even number of bridges, then it is possible to start from one of A or B and end up at the other, after using each bridge exactly once. In honour of these insightful discoveries, a network in which it is possible to come back to the starting point using each arc exactly once is said to have an **Euler walk**.

The approach Euler introduced of using representations in terms of nodes and arcs has proved useful in solving other puzzles as well, and we look at two such in this chapter.

THE JUGS PROBLEM

We saw in the first chapter that the solution to the problem about the three jugs of capacities 8, 5 and 3 litres can be thought of as trying to find a path in a network, where the starting state can be represented as [8,0,0] and the goal state as [4,4,0].

However, we will have to make some changes in the program we developed for finding a path between various cities before we can use it for the jugs problem. This is because there is an important difference between the structures of the two problems. In the cities problem, we knew explicitly that from city_1 there was a link to city_2, or that from city_3 there was a link to city_5, and so on. But in the jugs problem, the knowledge about which states can be reached from a given state is implicit, based on certain rules. For instance, from the starting state [8,0,0] we find out that the system can go to state [3,5,0] by using the additional knowledge that it is possible to fill up jug_b by pouring water from jug_a into jug_b, leaving 3 litres in jug_a and 5 in jug_b. Similarly, we use the knowledge that it is possible to fill jug_c from jug_a to deduce that it is possible to move to a state [5,0,3]. So we will have to supply the system with this knowledge, and ask it to use the knowledge to find out the possible next states.

What is the logic underlying the conclusion that it is possible to pour 5 litres from jug_a into jug_b when we start with 8 in jug_a and 0 in jug_b? We could say that it is a specific instance of the general rule that the quantity we can pour from jug_a to another is the lesser of two things:

(1) the quantity available in the jug from which we pour, jug_a.
(2) the empty space available in the jug into which we pour, jug_b.

Having determined the quantity which can be poured (provided it is greater than 0) we can determine the new state by respectively subtracting and adding this quantity to the quantities already in the jug from which we pour and into which we pour.

We can express this in Prolog by the following conditional clause:

```
pour([Q,jug_a,jug_b],[A,B,C],[A1,B1,C1]) :-
                    Free is 5-B,    lesser(A,Free,Q),
                         Q>0,
                         A1 is A-Q,
                         B1 is B+Q.
```

We can read this to say that pouring a quantity Q from jug_a to jug_b, changes the state from [A,B,C] to [A1,B1,C1], where

Q is the lesser of A and 5−B,
Q is greater than zero, and
A1 is A−Q and B1 is B+Q.

In writing this, we assume that it is possible to define a predicate 'lesser' which will take two numbers for the first two arguments, and give the smaller one as the answer, so that for example if we ask ?-lesser(8,5,What). We will get the response What=5. We will look at how to design clauses for lesser shortly. It is in fact a useful general technique in Prolog to assume that one can design any auxiliary or helping predicates which may be necessary, and design the main clause using such predicates, and then address the question of designing the auxiliary clauses. This is sometimes called a 'top down' approach to program design. Assuming we have such clauses for 'lesser', we see that if we ask:

```
?-pour([Q,jug_a,jug_b],[8,0,0],What).
```

the response we would get is:

What = [3,5,0].

This is because the system will bind the variable Free to 5−0, i.e. 5, and bind Q to 5 using lesser, and then bind A1 to 3 and B1 to 5.

Notice also that although Prolog tries to express knowledge in an English-like manner, there are some rules to be followed. One of these is that for arithmetic operations, like addition, subtraction, etc., the built-in operator 'is' has to be used. So, instead of expressing the new state as

[A−Q, B+Q, C],

we have to use A1 is A−Q and B1 is B+Q in the body of the conditional clause.

Next, let us consider how we can design clauses for 'lesser'. We can describe what is needed by saying that:

the lesser of First and Second is First,
if First is smaller than or the same as Second;
the lesser is Second,
if Second is less than or equal to First.

This translates into the Prolog clauses:

```
lesser(First,Second,First) :-
                    First=<Second.
lesser(First,Second,Second) :-
                    Second<First.
```

What if First equals Second? The first clause would be triggered, since Prolog starts searching the knowledge base from the top, and we would get the right answer.

We now have clauses which will work out what state changes are possible by pouring the appropriate quantity from jug_a to jug_b. We need five similar clauses for pouring from

jug_b to jug_a,
jug_a to jug_c,
jug_c to jug_a,
jug_b to jug_c, and
jug_c to jug_b.

All these would be similar clauses, and are given in a listing at the end of the chapter.

The program for this problem is similar to the problem of finding a path when there are two-way links rather than one-way links, because one can pour from jug_a to jug_b as well as from jug_b to jug_a. So we need to use the same technique to make sure that we do not go round in circles (pouring 5 litres from jug_a to jug_b and back to jug_a and back to jug_b . . .) by using a check to see that the Next state one goes to is *not* a state already visited.

It will also be useful to use another 'accumulator' list to gather up all the various transfers made, such as [5,jug_a,jug_b] so that at the end we can display the total sequence of transfers made as well as the sequence of states passed through. Let us use the predicate 'sequence' as an analogue of 'route' in the previous chapter. The clauses we need would be:

```
sequence(Start,End,Answer,Moves) :-
        sequencea(Start,End,Answer,Moves,[Start],[ ]).
sequencea(Same,Same,Answer,Moves,Answer,Moves).
sequencea(Start,Finish,Answer,Moves,Sofar,Movesacc) :-
        pour(Action,Start,Next),
    not(member(Next,Sofar)),
    sequencea(Next,Finish,Answer,Moves,[Next|Sofar],
        [Action|Movesacc]).
```

We can invoke the program with:

```
?-sequence([8,0,0],[4,4,0],States,Transfers).
```

and finally we will get a response showing a list of States visited, and a list of Transfers made. In order to make it easier for the user, a starting clause has been added as follows:

```
jugs :- write('what is the starting state ?'),
    read(Start), nl,
    write('what is the goal state?'),
    read(Finish),
    sequence(Start,Finish,Answer,Moves),
write('the successive states of the system are:'),
     nl,wl(Answer),nl,nl,
writc('thc scrics of transfcrs to bc madc arc:'),
     nl, wl(Moves).
```

This interacts with the user by asking for the starting state, which the user can supply as [8,0,0], and the goal state, which the user will say is [4,0,0]. Using these the program invokes the sequence clause, and finally provides a response in the form:

```
the successive states of the system are:
[8,0,0]
[3,5,0]
[0,5,3]
[5,0,3]
[5,3,0]
[2,3,3]
[2,5,1]
[7,0,1]
[7,1,0]
[4,1,3]
[4,4,0]

the series of transfers to be made are:

[5,jug_a,to,jug_b]
[3,jug_a,to,jug_c]
[5,jug_b,to,jug_a]
[3,jug_c,to,jug_b]
[3,jug_a,to,jug_c]
[2,jug_c,to,jug_b]
[5,jug_b,to,jug_a]
[1,jug_c,to,jug_b]
[3,jug_a,to,jug_c]
[3,jug_c,to,jug_b]
```

By allowing the user to prescribe both the start state and the end state, the program is flexible enough to allow us to find out the moves for getting from [8,0,0] to any finishing state, not merely [4,4,0].One of the exercises at the end of the chapter asks you to consider how the program can be made even more flexible, by allowing jugs of different capacities to be used. We saw when we analysed the program for routes that the successive states or nodes visited by the system are added at the head of the list, so that finally the list has the last state as the first element, i.e. it is in reverse order. The same thing happens here in the Sofar and Movesacc lists. One might like to reverse the lists before displaying them. A program for reversing lists, using the same technique of using an auxiliary predicate and an additional argument as an accumulator, has also been added to the listing at the end of the chapter. We can use this to modify our sequence clause to read as below:

```
sequencea(Same,Same,Answer,Moves,Revans,Revmoves):-
        sequencea(Start,Finish,Revans,Revmoves,[Start],[ ]),
            reverse(Revans,Answer),
              reverse(Revmoves,Moves).
```

It is because of these modifications made to the program that the sequence of states visited and the transfers made are shown in the correct order and not in reverse order in the output shown above.

MISSIONARIES AND CANNIBALS

Another famous puzzle to test one's wits is the missionaries and cannibals problem. Three missionaries and three cannibals are travelling together, and arrive at the left bank of a river. There is a boat available for crossing the river, but it can carry at most two persons, so obviously more than one trip is needed. There is a further complication in that at no stage must the missionaries be outnumbered by the cannibals on either bank of the river. (This is in older versions of the puzzle, where if the missionaries are outnumbered, they will be eaten up. In newer versions, the cannibals must not be outnumbered, otherwise they will be converted.)

The transition network method of representation is again very useful. We can represent the initial state by a list, showing as the first two elements the bank where the boat is and the bank to which it is headed, which would be l(eft) and r(ight) respectively to start with. The middle two elements would be the number of missionaries and cannibals on the side where the boat is, and the last two elements the numbers on the bank to which the boat is headed. So, the starting list would be [l,r,3,3,0,0].

We can identify the goal state as a situation when all of them are on the right bank. This would be [r,l,3,3,0,0], with the boat on the right bank, and all missionaries and cannibals on that bank too.

From the starting state, we see that the following changes of state are possible:

        one cannibal goes in the boat,
or   one missionary and one cannibal go over,
or   two cannibals go over.

It is not possible for one missionary or two missionaries to go in the boat, since that means there will be more cannibals than missionaries on the left bank.

We can show these changes of state diagrammatically as in Fig. 2.3. As we are using the third and fourth elements of the list to show the number of missionaries and cannibals on the bank where the boat is, when one cannibal rows over tothe right, these two elements of the list are 0,1, and similarly for the other two possibilities.

Clearly, the first branch of the network is not much use, for the cannibal has to bring the boat back, and that means the problem will be at the initial state again. The other two branches allow the possibilities of one missionary bringing the boat back, and of one cannibal bringing the boat back, respectively, leading to changes of state as in Fig. 2.4.

In the analysis above we have kept note of whether the next journey of the boat has to be from left to right, or right to left. When we develop the program we will have to provide for the current journey to be in the opposite direction to the previous journey.

The network could be developed further to find the path which leads to the goal state of [r,l,3,3,0,0], with the boat on the right bank.

From our knowledge of the jugs problem we see that we can use the same approach of providing the Prolog system with the rules for deciding which state changes are possible. As in the jugs problem, we should first specify clearly the relevant pieces of knowledge and logic we want the program to use. The key factors one has to keep in mind in deciding what journeys are possible are:

(1) the capacity of the boat. It can carry:

	one missionary,
or	one cannibal,
or	one missionary and one cannibal,
or	two missionaries
or	two cannibals.

In none of these cases is there any problem of missionaries being outnumbered by cannibals in the boat itself. We can provide unconditional Prolog clauses as below to use this knowledge, where the first argument of 'boattakes' shows the number of missionaries, and the second argument the number of cannibals:

```
boattakes(1,0).
boattakes(0,1).
boattakes(1,1).
boattakes(2,0).
boattakes(0,2).
```

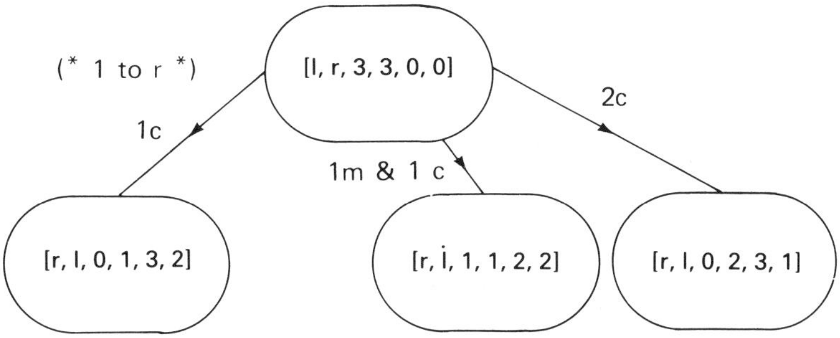

Fig. 2.3.

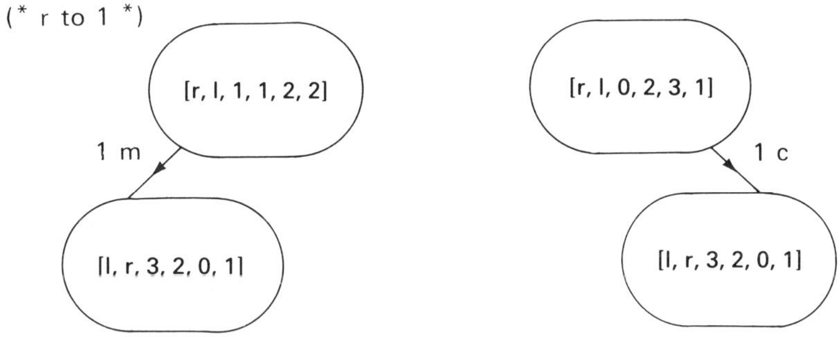

Fig. 2.4.

(2) We can reduce the number of people on the bank from which the boat leaves, and increase the number of people on the bank it is headed for, to compute the new state based on the old state and the number of people in the boat. But we will have to check that there are at least as many missionaries as cannibals on each bank, except for the case where there are only cannibals and no missionaries at all. We can represent this in Prolog as follows:

```
ok(0,Cann).
ok(Miss,Cann) :-
Miss>=Cann.
```

(3) Now, corresponding to the rules we developed for 'pour', we can give the system a conditional clause with predicate 'ferry' as follows:

```
/*      Mf is for Missionaries where boat is coming from
        Cf is for where boat is starting from
        Mt is for where boat is going to
        Ct is for cannibals where boat is going to
        Mb is for missonaries in the boat
        Cb is for cannibals in the boat          */

ferry([From,To,Mf,Cf,Mt,Ct],[To,From,Mfnew,Cfnew,Mtnew,Ctnew],
            [m,Mb,c,Cb]) :-
                  boattakes(Mb,Cb),
                  Mb=<Mf,
                  Cb=<Cf,
                  Mfnew is Mt+Mb,
                  Cfnew is Ct+Cb,
                        ok(Mfnew,Cfnew),
                  Mtnew is Mf-Mb,
                  Ctnew is Cf-Cb,
                        ok(Mtnew,Ctnew).
```

Let us see what happens if we ask a question:

```
?- ferry([left,right,3,3,0,0],Next,Boat).
```

We get the first response as:

```
Next=[right,left,0,1,3,2]
Boat=[m,0,c,1];
```

which corresponds to one cannibal rowing over from left to right.

If we ask for an alternative solution, we get:

```
Next=[right,left,1,1,2,2]
Boat=[m,1,c,1];
```

i.e. one missionary and one cannibal row over to the right bank. The system will also give us the third possibility:

```
Next=[right,left,0,2,3,1]
Boat=[m,0,c,2];
```

(4) We need to use our ferry predicate recursively to find the Next state, and successive Next states until the final state of [right, left,3,3,0,0] is reached. As we saw earlier we have to keep track of where the boat is and ensure that after a journey from left to right, the next journey has to be right to left and vice versa. One way of coding this is shown below.

(a) The first clause, 'misscan', is just so that the user can invoke the program simply by typing in one word.

```
          misscan :-
journeysa([l,r,3,3,0,0],[r,l,3,3,0,0],Answer,[[l,to,r,3,3,0,0]],
                    [ ]).
```

(b) 'misscan' calls on the journeysa clauses, which we see are very similar to the sequencea clauses we used in the jugs program. When you look at the code below, you will notice that it is very similar to the code for the jugs problem, or for that matter the route problem. In the listing at the end of the chapter, I have added some lines of comment to the code to highlight the similarities.

```
        /*  boundary case for journeysa   */
journeysa(Finish,Finish,Answer,Answer,[H|Boatacc]) :-
                    reverse(Answer,Revans),
               write('the successive states are:'), n1,
                    wl(Revans), nl,nl,
          write('hit any key plus period to continue'),read(Y),nl,
                    reverse([H|Boatacc],Revboat),
               write('the successive journeys are:'),nl,
                    wl(Revboat),nl,
               write('hit any key plus period'),nl,
                    read(X).
        /* recursive clause for journeysa, initially invoked by misscan with
        From bound to l for left, and To bound to r for right */

         journeysa([From,To|Start],Finish,Answer,Sofar,Boatacc) :-
                    ferry([From,To|Start],[To,From,P,Q,R,S],Boat),
                    not(member([To,to,From,P,Q,R,S],Sofar)),
                    journeysa([To,From,P,Q,R,S],Finish,Answer,
                         [[To,to,From,P,Q,R,S]|Sofar],
                    [[From,to,To|Boat]|Boatacc]).
               wl([ ]).
```

```
wl([H|T]) :-
  write(H),nl,
  wl(T).
```

When we invoke the program with misscan, the response we get on the screen is as below:

```
the successive states are:

    [l,to,r,3,3,0,0]
    [r,to,l,1,1,2,2]
    [l,to,r,3,2,0,1]
    [r,to,l,0,3,3,0]
    [l,to,r,3,1,0,2]
    [r,to,l,2,2,1,1]
    [l,to,r,2,2,1,1]
    [r,to,l,3,1,0,2]
    [l,to,r,0,3,3,0]
    [r,to,l,3,2,0,1]
    [l,to,r,1,1,2,2]
    [r,to,l,3,3,0,0]

hit any key plus period to continue
the successive journeys are:

    [l,to,r,m,1,c,1]
    [r,to,l,m,1,c,0]
    [l,to,r,m,0,c,2]
    [r,to,l,m,0,c,1]
    [l,to,r,m,2,c,0]
    [r,to,l,m,1,c,1]
    [l,to,r,m,2,c,0]
    [r,to,l,m,0,c,1]
    [l,to,r,m,0,c,2]
    [r,to,l,m,1,c,0]
    [l,to,r,m,1,c,1]

hit any key plus period.
```

SUMMARY

In this chapter, we have seen how the general concept of finding a path through a network provides a useful approach to solving puzzles. In the succeding chapters, we will look at how formulating a problem in terms of a network of states and searching for paths in the network has various other useful applications.

EXERCISE

In the jugs program, the starting clause 'jugs' allows the user to specify any start state and any end state. So the sequence of pouring required to start with [8,0,0] and end up with [6,2,0] can also be found out with this program. The program could be made even more general if we provide for three jugs of any capacities, not just 8, 5, and 3. After looking at the listing in Appendix 1, you may like to think of some way to modify the code to enable this to be done.

APPENDIX 1: LISTING OF THE JUGS PROGRAM

```
jugs :- write('what is the starting state ?'),
            read(Start), nl,
        write('what is the goal state?'),
            read(Finish),
    sequence(Start,Finish,Answer,Moves),
        write('the successive states of the system are:'),
            nl,wl(Answer),nl,nl,
        write('the series of transfers to be made are:'),
            nl, wl(Moves).

sequence(Start,End,Answer,Moves) :-
            sequencea(Start,End,Answer,Moves,[Start],[ ]).

sequencea(Same,Same,Answer,Moves,Revans,Revmoves):-
            reverse(Revans,Answer),
        reverse(Revmoves,Moves).

sequencea(Start,Finish,Answer,Moves,Sofar,Movesacc):-
            pour(Action,Start,Next),
        not(member(Next,Sofar)),
sequencea(Next,Finish,Answer,Moves,[Next|Sofar],
                                    [Action|Movesacc]).

    lesser(First,Second,First) :-
                    First=<Second.
    lesser(First,Second,Second) :-
                    Second<First.
```

```
pour([Q,jug_a,to,jug_b],[A,B,C],[A1,B1,C]) :-
                                   Free is 5-B,
                              lesser(A,Free,Q),
                                   Q>0,
                                   A1 is A-Q,
                                   B1 is B+Q.

pour([Q,jug_a,to,],[A,B,C],[A1,B,C1]) :-
                              Free is 3-C,
                         lesser(A,Free,Q),
                              Q>0,
                              A1 is A-Q,
                              C1 is C+Q.

pour([Q,jug_b,to,jug_a],[A,B,C],[A1,B1,C]) :-
                                   Free is 8-A,
                              lesser(B,Free,Q),
                                   Q>0,
                                   A1 is A+Q,
                                   B1 is B-Q.

pour([Q,,to,jug_a],[A,B,C],[A1,B,C1]) :-
                              Free is 8-A,
                         lesser(C,Free,Q),
                              Q>0,
                              A1 is A+Q,
                         Cl is C-Q.

pour(Q,jug_b,to,],[A,B,C],[A,B1,C1]) :-
                              Free is 3-C,
                         lesser(B,Free,Q),
                              Q>0,
                              B1 is B-Q,
                           C1 is C+Q.

pour(Q,,to,jug_b],[A,B,C],[A,B1,C1]) :-
                              Free is 5-B,
                         lesser(C,Free,Q),
                              Q>0,
                              C1 is C-Q,
                              B1 is B+Q.
```

```
/* wl is for writing out a list in the format of each item on one line */

wl([ ]).
wl([H|T]) :- write(H),nl, wl(T).

reverse(L,Answer) :-
            reversea(L,Answer,[ ]).

      reversea[H|T],Ans,Sofar) :-
            reversea(T,Ans,[H|Sofar]).
      reversea([ ],Answer,Answer).
```

APPENDIX 2: LISTING OF MISSIONARIES AND CANNIBALS PROGRAM

```
boattakes(1,0).
boattakes(0,1).
boattakes(1,1).
boattakes(2,0).
boattakes(0,2).

ok(0,Cann).
ok(Miss,Cann) :-
        Miss>=Cann.

/*   Mf is for Missionaries where boat is coming from
     Cf is for where boat is starting from
     Mt is for where boat is going to
     Ct is for cannibals where boat is going to
     Mb is for missionaries in the boat
     Cb is for cannibals in the boat          */

ferry([From,To,Mf,Cf,Mt,Ct],[To,From,Mfnew,Cfnew,Mtnew,Ctnew],
                              [m,Mb,c,Cb]) :-
                                 boattakes(Mb,Cb),
                                 Mb=<Mf,
                                 Cb=<Cf,
                                 Mfnew is Mt+Mb,
                                 Cfnew is Ct+Cb,
                                   ok(Mfnew,Cfnew),
                                 Mtnew is Mf-Mb,
                                 Ctnew is Cf-Cb,
```

```
                                ok(Mtnew,Ctnew).

        misscan :-

        journeysa([l,r,3,3,0,0],[r,l,3,3,0,0],Answer,[[l,to,r,3,3,0,0]], [ ]).
        /*   boundary case for journeysa   */

        journeysa(Finish,Finish,Answer,Answer,[H|Boatacc]) :-
                        reverse(Answer,Revans),
                    write('the successive states are:'), nl,
                        wl(Revans), nl,nl,
                write('hit any key plus period to continue'),read(Y),nl,
                        reverse([H|Boatacc],Revboat),
                    write('the successive journeys are:'),nl,

                        wl(Revboat),nl,
                    write('hit any key plus period'),nl,
                        read(X).
```

/* recursive clause for journeysa, initially invoked by misscan with From bound to l for left, and To bound to r for right */

/* This is similar to the jugs sequencea recursive clause in the following respects:

It uses an accumulator Sofar for all the previous states, to make sure that there is no going round in circles. The variable Boatacc keeps building up a list of the various trips made, so that at the end the non-recursive clause for journeysa can print out this list */

```
journeysa([From,To|Start],Finish,Answer,Sofar,Boatacc) :-
          ferry([From,To|Start],[To,From,P,Q,R,S],Boat),
          not(member([To,to,From,P,Q,R,S],Sofar)),
          journeysa([To,From,P,Q,R,S],Finish,Answer,
               [[To,to,From,P,Q,R,S]|Sofar],
          [[From,to,To|Boat]Boatacc]).
```

/* this is used in the non-recursive journeys clause to print out a list of items in the form one item on each line */

```
      wl([ ]).
      wl([H|T]) :-
```

```
                write(H),nl,
                wl(T).

reverse(Given,Answer) :-
                reversea(Given,Answer,[ ]).

reversea([H|T],Answer,Sofar) :-
                reversea(T,Answer,[H|Sofar]).

reversea([ ],Answer,Answer).
```

3

Networks and dynamic programming

PREVIEW

In the first chapter, we looked at the basic terminology of transition networks: nodes, arcs, transitions between nodes and paths. We also discussed the idea of defining a program recursively, as applied in particular to the task of tracing a path in a network. In the next chapter, we saw how these concepts come in useful in solving some classic puzzles. We will now see in this chapter how these basic ideas lead to some useful applications. However, before we go on to this, I would like to say something more about the terminology used in this area.

WHAT IS GRAPH THEORY?

The ideas we have seen in the last two chapters are fundamentally the expression of a certain point of view, or approach, in which situations are analysed in terms of **states** of a system, and certain possible **transitions** between pairs of states. These relationships between pairs of states lead quite naturally, as we have seen, into considering collections or networks of states. Further, the solutions to various problems can be identified with tracing paths in the networks. Instead of thinking of the nodes as states, and the relationships between nodes as transitions, one can alternatively think of the same structure in more abstract terms. The nodes can be thought of as just any set of objects or entities, and the arcs or links can be thought of as just a selection of some or all of the pairs of nodes. The pairs which are satisfied are those which have some relationship, which is denoted by an arc between them. This more abstract formulation of what are essentially the same ideas is known as **graph theory**, a name coined by the nineteenth century mathematician Cayley. A finite graph is defined as consisting of:

(a) a finite set of entities called the nodes of the graph.

(b) a finite set of pairs of these nodes,called the arcs or links of the graph.

Clearly, this definition too leads to a diagrammatic representation which is the same as the one for the transition networks we have considered, and this is called a **graph**.The only difference is that we think of the arc between two nodes as just some undefined relationship between the nodes rather than more concretely as a transition. In this usage of the word graph,of course, it means something different from the more common use of graph to indicate a line showing, for example, the year on the horizontal axis and the rate of inflation on the vertical axis. Incidentally, graph theory can also be extended to consider cases where the number of nodes or of arcs or of both is not finite, but we will not need to consider such graphs for the applications discussed in this book.

Also, when we consider the set of pairs of nodes, we can make a distinction whether or not the order of the pair matters, that is whether the pair (a,b) of nodes is the same or is different from the pair (b,a). In diagrammatic terms, saying (a,b) and (b,a) are the same is equivalent to saying that there is no direction associated with the arc. In that case, the arc is called an **undirected arc**, and if the whole graph consists of undirected arcs it is called an **undirected graph** or just **graph**. On the other hand, treating (a,b) as different from (b,a) is equivalent to saying that the arc (a,b) is associated with an arrow from a to b. There may or may not be an arc going the other way from b to a, and the distance or other label associated with the arc from b to a could be different from the one associated with the arc from a to b. In that case, the arcs are called **directed arcs**, and if the whole graph consists of directed arcs it is called a **directed graph** or **digraph**. Clearly, we can treat undirected graphs as a special case of directed graphs, where there are always arrows both ways, and the distance or whatever associated with the arc is the same both ways. The terminology associated with graph theory is far from standardized. Nodes are also called vertices or points, and arcs are also called links or edges. Also, sometimes a distinction is made that if there is a label like a distance or action or something else associated with every arc, this is called the **weight** of the arc, and the graph is called a weighted graph (or digraph) or a **network**, but if just the relationship is shown, it is just a graph (or digraph). In this book, we will generally be using the term transition network or directed graph.

SHORTEST PATH AND DYNAMIC PROGRAMMING

We looked at two key concepts in the first chapter, the concept of tracing a path, and the concept of recursively specifying or describing a problem or its solution. We will now look at a problem-solving technique known as **dynamic**

programming, which applies these concepts. As we will see, the technique can be used most straightforwardly for finding the shortest path in a network. But it is even more useful, for it turns out that many other problems which at first sight do not seem to be about finding a path can be represented in a way which converts them into a path problem.

We saw that it was possible to design a program which would trace all possible paths in a network from a given starting point to a given destination, and also give us the length of the path (the 'routea' program we saw in Chapter 1). It would seem to be a straightforward matter to find all the paths and find which one is the shortest. However, while this is a possible solution in principle, in practice it leads to complications. Suppose from a given starting point there are possible arcs to three nodes aa,bb and cc, i.e. at what one might call the first **stage** there are three possible paths. (I have designated the nodes as aa,bb,cc, etc., because when we want to represent it in Prolog, we need names starting with a lower case letter, and using a single letter 'a' might not make it stand out that we are referring to a node.) Now from aa there may be arcs to say three nodes, and from bb to five nodes and from cc to two nodes, so we have ten possible paths for the first two stages, that is when we consider paths whose length is two arcs from the start. Again, at the next stage there may be three arcs from each of the nodes reached in the second stage and so on, so we see the number of possible paths multiplies very rapidly at each stage. In a network which involves 10, 15, 20 . . . stages the number of paths may become too large for all of them to be evaluated in order to find the shortest path. The method of dynamic programming is a technique which avoids having to evaluate all the possible paths, as will become clear when we look at the method.

The key concept on which the method is built is this. Suppose we want to find the shortest path from a given starting point to a given destination, and suppose from the starting point, or node 'start', there are arcs to three other nodes aa, bb and cc. Then it must be true that the shortest path we are looking for is the smallest of the following:

(1) the distance from start to aa plus the length of shortest path from aa to the destination.
(2) distance from start to bb plus the length of shortest path from bb to the destination.
(3) distance from start to c plus the length of shortest path from cc to the destination.

. . . and so on, if there are more nodes dd,ee . . . to which there is a direct arc from start. Each of these is called an **immediate successor** of node start, and start is called an **immediate predecessor** of these nodes.

We see that this specification of the problem and its solution shows the characteristic feature of a recursive description, that we use the concept of shortest path we are defining as part of the definition, but with some change in the arguments (instead of start and destination, we have aa and destination, bb and destination, etc.). Or, to look at it another way, we have represented the problem in a way where it involves a subproblem of the same type.

As an illustration, suppose the start node has arcs of lengths 5, 4 and 7 leading to aa, bb and cc, as in Fig. 3.1. We have reduced the original problem

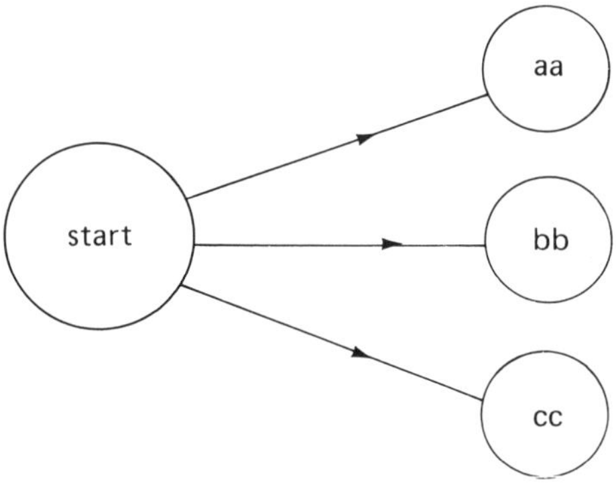

Fig. 3.1.

to three subproblems of the same type, for we now need to find the length of shortest paths from aa, bb and cc to destination, so that we can add 5, 4 and 7 respectively to these lengths and find the smallest of the three sums. But our definition implies that in order to find the shortest path from, for instance, aa,to the destination, we have to find the immediate successors of aa, let us say to dd and ee, with lengths of 2 and 7, as in Fig. 3.2, and then we have to solve the new subproblems of finding the shortest paths from dd and ee to the destination.

Once again, the subproblems of finding the shortest paths from dd and ee to the destination will lead us to further subproblems of finding the shortest path from the nodes to which dd has a direct arc, and so on . . . . However, at some stage we are going to reach a situation as in Fig. 3.3. This is a situation where from pp, qq and rr there are direct links *only* to the destination (i.e. their only immediate successor is the destination node), and it is clear that the shortest paths from pp, qq or rr to the destination are just the lengths of the links pp to destination, qq to destination and rr to destination. So this is the

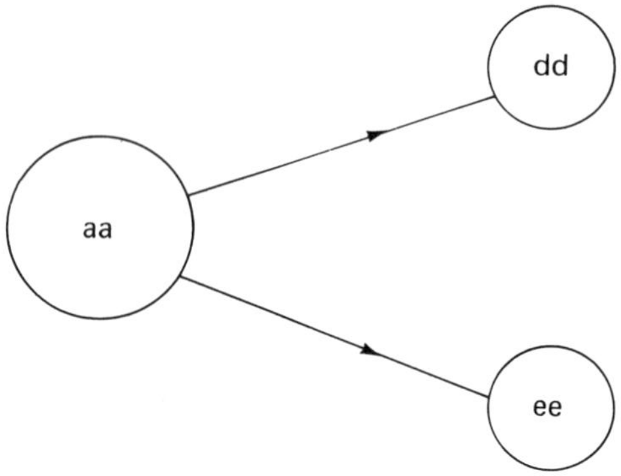

Fig. 3.2.

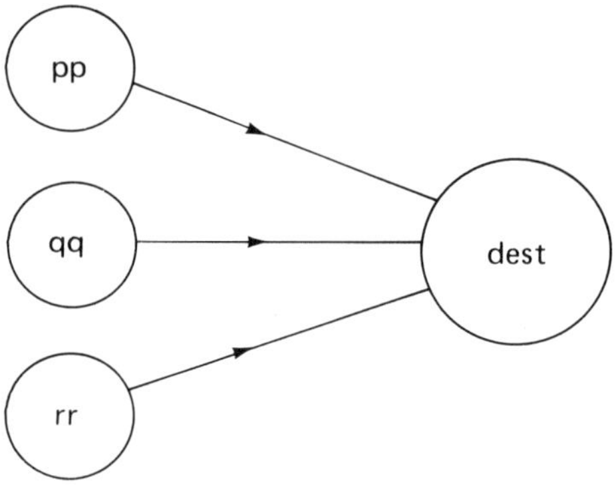

Fig. 3.3.

boundary or non-recursive case. This simple solution for the non-recursive case will give us the information we want for finding the shortest path from the immediate predecessors of pp, qq and rr, and that in turn will give us the answer for the immediate predecessors of those immediate predecessors, and so on till we come back to the stage where the immediate predecessor is the start node. We could have carried the recursion one step further, and said that the shortest path from pp to destination is the smallest of the sums got by adding up the distances of the links from pp to the destination to the shortest distance from the destination to the destination. In that case, the non-recursive case would be when we want to find the distance from the destination to the destination, which is obviously zero.

To illustrate how this works, let us consider an example of a network with seven nodes, start, aa, bb, cc, dd, ee and dest, as in Fig. 3.4.

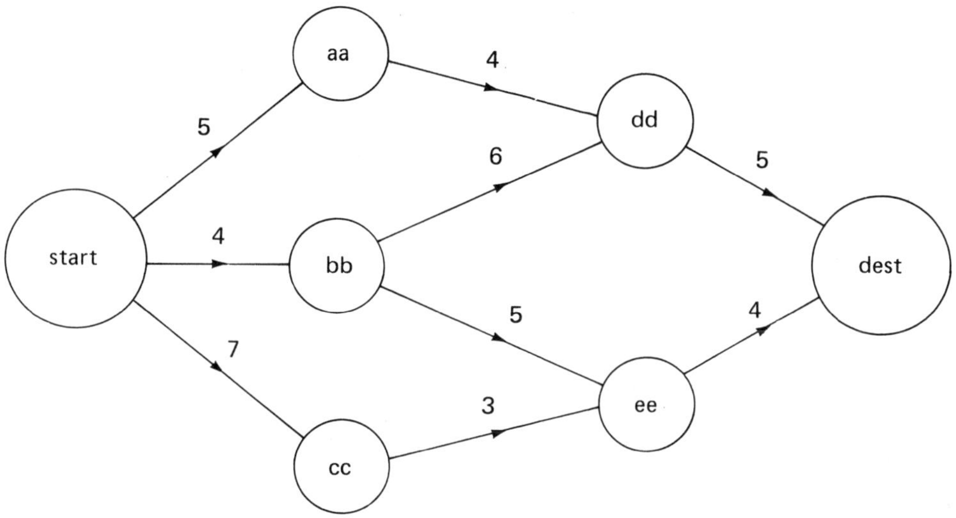

Fig. 3.4.

Drawing a diagram showing the arcs and the distances may not be suitable for larger problems, so let us look at an alternative way of representing the relevant information in a table. We know that we first want to solve the problem for the non-recursive case, i.e. for the nodes with direct arcs only to the destination, so we start with the information only for those arcs:

Table for stage 3

From	To dest
dd	5
ee	4

In the case of dd and ee, which have direct arcs only to the destination, we know the shortest path from dd to the destination is 5 and from ee to the destination is 4. In terms of the terminology of stages we looked at earlier, we have solved the problem for stage 3.

Now we start using the recursion approach and look at the nodes which have direct arcs to dd or ee or both, i.e. we are looking at nodes in stage 2. The information for this is as follows:

Table for stage 2

From	To	
	dd	ee
aa	4	x
bb	6	5
cc	x	3

An x indicates there is no direct arc, for example from aa to ee.

From the row for cc we see that cc has a link only to ee, so the shortest path from cc must be the smaller of:

(1) 3 + length of shortest path from ee (=4),i.e. 7 through ee.
(2) there is no (2) since in this particular example, cc has no other immediate successor.

Similarly, from the row for bb we see that bb has direct links to two nodes dd and ee, so the shortest path from bb must be the smaller of:

(1) 6 + shortest path from dd (=5), i.e. 11
(2) 5 + shortest path from ee (=4), i.e. 9

So the shortest path from bb is 9 through ee.

Next, aa has a direct link only to dd, we can say by applying the same principles, that the shortest path from aa is 4+5=9 through dd.

We repeat the same procedure and look for nodes which have direct arcs only to one or more of aa, bb and cc, i.e. we are now at stage 1. This information is as follows:

Table for stage 1

From	To		
	aa	bb	cc
start	5	4	7

We can see that the shortest path from start, which the table above tells us has direct links only to aa, bb, and cc must be the smallest of:

(1) 5+9=14 through aa
(2) 4+9=13 through bb
(3) 7+7=14 through cc

So the shortest path is 13 through bb, and we know the shortest path from bb was through ee, and we know the shortest path from ee was direct to dest, so we know the shortest path from start to destination is:

start ⟶ bb ⟶ ee ⟶ dest.

We could also have represented the information in a combined table or matrix form of representing a network, as follows:

From	To						
	start	aa	bb	cc	dd	ee	dest
start	0	5	4	7	x	x	x
aa		0	x	x	4	x	x
bb			0	x	6	5	x
cc				0	x	3	x
dd					0	x	5
ee						0	4
dest							0

The table shows 0 as the distance from any node to itself, and has an 'x' when there is no direct link. Also, we are considering a network in which there is a clear progress from stage to stage — from start, which we can think of as stage 0, we go to any one of aa, bb and cc in stage 1, and from there to any one of dd or ee in stage 2, and from there to dest in stage 3. So, there is no question of looping back, and so there are no entries below the diagonal of 0s for backward links. We will look later at applications where it is useful to have the whole table, but here it is handy to use subtables as we have done above.

We see that this recursive method of calculating the shortest path gives us the advantage of allowing us to systematically make sure that we have taken into account all possible paths. In a really large network, it would be quite easy to overlook possible paths. However, even more important is the reduction in calculation involved. To get an idea of this, let us say we have a network in which in the first stage one could go from start to any one of 10 possible nodes. In the second stage there may be another 10 possible nodes one could go to from each of the first stage nodes, and similarly in the third stage there may be another 10 nodes to which one could get from each of the second stage nodes. Finally, one could get from each of these third stage nodes to the destination in the fourth stage, as shown in Fig. 3.5.

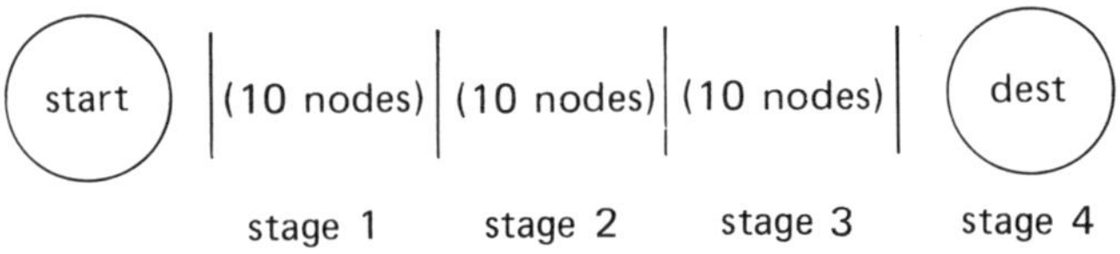

Fig. 3.5.

We see that there are $10\times10\times10=1000$ possible paths, each consisting of four arcs. Calculating the length of each of these paths requires adding up four numbers, which we can think of as equivalent to adding up pairs of numbers three times, by first adding the first two numbers to get a result, next adding the third number to this result, and finally adding the fourth number to get the length of the path. So, the total computation is the addition of 3000 pairs of numbers, plus 999 comparisons to find the smallest number out of the lengths of the 1000 paths. On the other hand, in using the dynamic programming method, we would start at stage 3 and get the shortest routes for the nodes connected directly to the destination. Then we would carry out additions of ten pairs of numbers, and nine comparisons of the ten totals we get, to get the shortest route from each of the stage 2 nodes to the destination. So, for all 10 stage 2 nodes, we would have a total of 100 additions, and 90 comparisons. Similarly, to get the shortest routes for the stage 1 nodes, we would carry out another 100 additions and 90 comparisons. Finally, we would carry out 10 additions and nine comparisons to get the shortest route for the start node. So we see the amount of computation is significantly reduced: 210 additions and 189 comparisons, as contrasted with 3000 additions and 999 comparisons. The advantage becomes greater, the more stages and the more nodes. For example, if there is one more stage with 10 nodes, in between the present stage 3 and the destination, with the new stage as stage 4 and the destination as stage 5, the number of possible paths between start and destination now becomes 10 000, and each path consists of five arcs, requiring the addition of four pairs of numbers. So we now need 40 000 additions of pairs of numbers and 9999 comparisons to find the smallest of the 10 000 sums to find the shortest of the 10 000 paths. The addition of a new stage with 10 nodes has increased the amount of arithmetic needed dramatically, with the number of additions going up much more than 10 times, and the number of comparisons going up about 10 times. On the other hand, if we use the dynamic programming method, it means now that we will get the shortest distance from the new stage 4 to the destination directly, but we we will have to calculate the shortest distance for each stage 3 node as follows: we will have to add the distance from that node to each of the 10 stage 4 nodes to the distance from that stage 4 node to the destination, and then pick the smallest number. So, we have 10 more additions and nine more comparisons for each stage 3 node, involving extra computation of 100 additions and 90 comparisons. Once we have calculated the shortest path from each stage 3 node, the calculations for the stage 2 and stage 1 nodes and for the start node remain the same. We see the increase in computation as a result of adding one more intermediate stage of 10 nodes to the previous intermediate stages stage 2 and stage 3 is just the 100 additions and 90 comparisons required by intermediate nodes, and is much much less than if we try to evaluate all the paths directly. The work required in the latter method increases very rapidly because of the number of paths increasing multiplicatively, whereas the dynamic programming method requires only a modest increase in the number of computations involved.

Now let us see how we could write a Prolog program which would carry out the computations. We first need to represent the following information about each node:

what nodes does it have a direct link to?
what are the distances of these links?

A convenient way of putting this in the knowledge base would be as below:

```
/* CLAUSES SHOWING THE STRUCTURE OF THE NETWORK OR RELATIONSHIPS BETWEEN THE VARIOUS NODES */

follow(start,[aa,bb,cc],[5,4,7]).
follow(aa,[dd],[4]).
follow(bb,[dd,ee],[6,5]).
follow(cc,[ee],[3]).
follow(dd,[dest],[5]).
follow(ee,[dest],[4]).
```

The first clause tells us that start is followed, or has direct arcs connecting it to, the nodes aa, bb and cc, with distances of 5, 4 and 7 respectively, and similarly for the other clauses. Now, let us look at the non-recursive case, of the nodes whose only immediate successor is the destination node:

```
/ *SP CLAUSE 1 */
/* NON-RECURSIVE BOUNDARY CASE FOR FINDING
   SHORTEST PATH */
      sp(Start,Finish,Length):-
          follow(Start,[Finish],[Length]).
```

We can see that a query like:

```
?-sp(ee,dest,Shortest).
```

will match and trigger this clause and would lead to the response:

```
Shortest=4
```

because the condition in the body of the rule:

```
follow(ee,[dest],4).
```

is satisfiable from the knowledge base.

Now let us look at a recursive clause for nodes other than those directly connected to the destination node.

```
/* SP CLAUSE 2 */
/* RECURSIVE RULE */
/* This rule finds all the successors, and then invokes the clause
'best' to find out the shortest route by finding the shortest of the
routes through each of the successors */
        sp(Start,Finish,Length) :-
            follow(Start,Allsucc,Distances),
            best(Allsucc,Distances,Finish,Length,[]).

/* BEST CLAUSE 1 */
        best([H1|T1],[H2|T2],Finish,Answer,Choices):-
            sp(H1,Finish,L1),
            D1 is H2 + L1,
          best(T1,T2,Finish,Answer,[D1|Choices]).

/* BEST CLAUSE 2 */
best([],[],Finish,Answer,Choices) :-
                smallest(Choices,Answer).
/* THIS IS AN AUXILIARY PROGRAM TO FIND THE
   SMALLEST OF A LIST OF NUMBERS */
/* SMALLEST CLAUSE 1 */
        smallest([Only],Only).
/* SMALLEST CLAUSE 2 */
        smallest([H1,H2|T],Rcsult):-
            H1>=H2,
              smallest([H2|T],Result).
/* SMALLEST CLAUSE 3 */
        smallest([H1,H2|T],Result):-
            H1<H2,
              smallest([H1|T],Result).
```

To use this program, one would ask:

```
?- sp(start,dest,Answer).
```

and the system would respond with Answer=13.

To see how this operates, let us first analyse what happens when we ask:

?-sp(bb,dest,Answer).

As we know, solving this subproblem of the shortest path from bb to the destination is in fact a component of solving the whole problem.

As mentioned in the first chapter, any statement between the pair of markers /* and*/ is treated by Prolog systems as just comments, and I have added some comments, and also headed the various clauses as /* BEST CLAUSE 1 */ or /* SMALLEST CLAUSE 2 */ and so on, for convenience of reference in the discussion below. Let us look at the various steps (1), (2), (3) . . . involved in the process.

(1) Since Prolog searches the knowledge base from the top, it will first try to match the goal set by the user: ?-sp(bb,dest,Answer). with the SP CLAUSE 1. The body of this clause will make the system look for a match for follow(bb,[dest],Length]). and this will fail, because dest is *not* the only immediate successor of bb.
(2) the SP CLAUSE 2 will now be tried.We can see that the body of this clause will lead to the goal follow(Start,Allsucc,Distances). being tried. This can be matched in the knowledge base, which contains follow(bb,[dd,ee],[6,5]). Notice Start with an upper case letter is treated as a variable, distinct from the specific node 'start' entirely lower case. As a result of this matching, variable Allsucc will be bound to the value [dd,ee]. and variable Distances will be bound to the value [6,5]. Note also that the variable Answer in the user's question would have been matched with the variable Length in the clause, so when a value is found for the latter, it will be passed on to the former.
(3) Next the second goal in the SP CLAUSE 2 will be tried. This will be:

    best([dd,ee],[6,5],dest,Length,[]).

    (taking into account all the bindings of values to variables which have taken place up to this stage).
(4) When this goal is tried on the BEST CLAUSE 1, the head of this clause will match dd with H1, [ee] with T1, dest with Finish, 6 with H2, [5] with T2, and Choices with []. The program will now go on to the first goal in the body of the clause: sp(dd,dest,L1).
(5) This goal will succeeed using SP CLAUSE 1,because as required by the body of that rule, a match can be found for follow(dd,[dest],[L1]) with follow(dd,[dest],[5]) which is in the knowledge base. As a result, L1 will be bound to the value 5.

(6) The system will now go on to the next goal in the BEST CLAUSE 1 which it is working with (see (3) above) and will be able to compute D1 is H2+L1 , since it has bound H2 to 6 and L1 to 5, and so will bind D1 to 11. We can see that what has happened is that the distance from bb to dd (6) has been added to the distance from dd to dest (5) to find the total length 11.

(7) the next goal in the body of the BEST CLAUSE 1 will now be tried, as:

best([ee],[5],dest,Answer,[11]).

We can see that the distance 11 of bb to dest through ee has been inserted in the last list, which was empty to start with.

(8) By a process similar to that analysed in (3) to (6) above, we can see that the distance bb to ee and the distance ee to dest will also be added,giving 5+4, and this will be inserted in the list to give [9,11]. Now, there will be a call to:

best([],[],dest,Answer,[9,11]).

(9) But this time, we see the BEST CLAUSE 1 will not work, because there is no way the empty lists can be broken up into H1 and T1 and H2 and T2 respectively. So the system will go on and try BEST CLAUSE 2.

(10) The head of this clause can be matched, and will lead to the only goal in the body of the clause: smallest([9,11],Answer).

(11) As one can guess, the answer will be 9, so that finally we will be given the answer that the shortest path from bb to dest is 9. But let us look at how the clauses for smallest carry out this task. AA The query or goal is smallest([9,11],Answer). and as always, the system will first try the topmost of the 'smallest' clauses, which is the one headed SMALLEST CLAUSE 1. This will fail, because it requires a list with only one element in it, [Only], whereas here there are two elements 9 and 11 in the list. So the next clause SMALLEST CLAUSE 2 will be tried. We see that H1 will be bound to 9, H2 to 11,and T to []. Answer will be identified with Result, so that whatever value is found for Result will be shared by Answer as well. The first part of the body of the clause is H1>=H2, which in this case is not true. So, the system cannot go ahead to consider the rest of the body of the clause, and will instead now go on to SMALLEST CLAUSE 3. In this case, the binding of H1 to 9 and H2 to 11 leads to H1<H2 in the body of the clause being true. So, the next goal in the body, namely:

smallest([H1|T],Result)

which is now equivalent to:

```
smallest([9],Result)
```

will be tried. We see that this new goal will triggger SMALLEST CLAUSE 1, because the list has only one element in it. So, the variables Only and Result will both be bound to 9, and ultimately, by passing back the values, we get the response Answer = 9, for our question:

```
?-sp(bb,dest, Answer).
```

It is now a question of extending this analysis to see what happens if we ask ?-sp(start,dest,Answer). Just as the shortest path from bb to dest was found to be 9, the shortest paths from the other immediate successors of start, namely aa and cc, will be found to be respectively 9 and 7, and after adding the distances of start from these three, the smallest will be 13 and this will be given as the response.

As we saw, the technique requires first finding the stage 3 shortest paths from dd and ee to dest, then using these results to find the shortest path at stage 2 from aa, bb and cc to dest, and then using these results to find the shortest path from start to dest. A convenient way to study how the program goes through these steps is to ask:

```
?- sp(From,dest,Answer).
```

and the program comes back with the following response:

```
?- sp(From,dest,Answer).
From = dd
Answer = 5 ->;

From = ee
Answer = 4 ->;

From = start
Answer = 13 ->;

From = aa
Answer = 9 ->;

From = bb
```

```
Answer = 9 ->;

From = cc Answer = 7 ->;

no

?-
```

We are able to see all the solutions to the subproblems as well as the solution to the original problem. (In the particular version of Prolog from which this output was captured, Arity Prolog, when a question has multiple answers the system produces an answer and waits for a ; to be typed as a signal that it should search for the next possible answer).

It would of course be even better if, in addition to giving the response 13, the actual path were also given. Readers may like to try their hand at making this extension. In the exercises at the end of this chapter, I have suggested the factors to be taken into consideration in making the extension, and the solutions to the exercises contain one possible way of coding this extension.

We have looked at the application of dynamic programming for finding the shortest path in terms of finding the shortest route from every node to the destination. So the non-recursive problem was for the last stage nodes, those which had arcs leading only to the destination, and we used this recursively for the earlier stages. A variant of this method is to find the shortest route from a starting point to every other node in the network. In this case, the simplest non-recursive case would be for the nodes which have arcs leading into them starting only from the starting node, i.e. in terms of what we have seen above, the stage 1 nodes. The information for the problem we solved earlier is reproduced below:

From	To						
	start	aa	bb	cc	dd	ee	dest
start	0	5	4	7	x	x	x
aa		0	x	x	4	x	x
bb			0	x	6	5	x
cc				0	x	3	x
dd					0	x	5
ee						0	4
dest							0

We start with stage 1, and can see immediately that for nodes aa, bb and cc, the shortest routes from start are 5, 4 and 7 respectively.

Now we need to use a recursive specification for nodes which are not immediate successors of start. We need tó do the following for each such node we are currently considering:

(a) consider in turn each immediate predecessor of the current node, take the shortest distance from start to that predecessor, and add the distance from the predecessor to the current node.
(b) take the smallest of these sums, which tells us the shortest distance from start to the current node.

So, now we take the nodes for which aa, bb, or cc are the immediate predecessors, i.e. dd and ee.

From	To	
	dd	ee
aa	4	x
bb	6	5
cc	x	3

The shorter distance from start to dd must be the smallest of:

(1) 4 + shortest distance to aa (=5) i.e. 9
(2) 6 + shortest distance to bb (=4) i.e. 10

So, shortest distance to dd is 9 and the route is from aa. Similarly, the shortest distance from start to ee must be the smallest of:

(1) 5 + shortest distance to bb (=4) i.e. 9
(3) 3 + shortest distance to cc (=7) i.e. 10

So, shortest distance to ee is 9 and the route is from bb.

Finally we come to stage 3, and by the same kind of argument, the shortest distance to dest must be the smallest of:

(1) 5 + shortest distance to dd (=9) i.e. 14
(2) 4 + shortest distance to ee (=9) i.e. 13

So, the shortest distance to dest is 13, through ee. And as before, since shortest distance to ee was through bb, and to bb was from start, we see the shortest route to dest is:

start ⟶ bb ⟶ ee ⟶ dest

which was the same result we found before.

In the exercises section of this chapter, there are suggestions on how a program can be written in Prolog to show the shortest path from start to any given node, and a listing is given in the solutions section.

MULTISTAGE DECISION PROCESSES

The concept of dynamic programming as we have seen it applied to the problem of finding a shortest path is capable of being generalized and applied in a much wider context. We can think of the technique as essentially concerned with a **multistage decision process**. In the shortest route problem, at stage 1, a decision had to be taken as to which of several alternative routes should be used. One can think of this in terms of starting at an initial state, and moving to one of several possible states. Having reached one of these states, in each of these a choice had to be made between various alternatives, each of which would lead to a different state. The process could go through any number of stages before reaching a final goal state, the destination. The designer of the dynamic programming approach, Richard Bellman, pointed out that a significant characteristic of such a multistage decision process is that if one finds the best or optimal sequence of decisions, it must be true that each subsequence of this optimal sequence must also be the best with reference to the starting and ending states of that subsequence. In terms of the shortest path, if the shortest path from start to destination passes through the sequence of states aa, cc, ee, gg, hh, qq, pp, then it must be true that the shortest path from cc to qq must be also cc, ee, gg, hh, qq and the shortest path from hh to the destination must be through hh, qq, pp and so on for any subsequence of the shortest path from start to destination. Now why is this true? We can establish it by assuming that it is not true, and showing that such an assumption leads to a contradiction. For example, if there were some shorter path between ee and hh other than ee, gg, hh, then we could modify our original shortest path, travelling from start to ee by the route aa, cc, ee, then taking the alternative shorter path to gg, and then going on from hh to qq, pp and destination, and thereby shortening the distance. But that is contradictory, for there cannot be any way of shortening the shortest path, so it shows that the shortest path between any two nodes on the shortest path between the start and destination is the same as the subpath between them on the shortest path between start and destination. This general principle, that each subsequence of a sequence of optimal decisions is itself optimal, is known as **Bellman's principle of optimality**. We have used the principle in recursively building up the shortest path from start to stage 4 by using first the shortest path from stage 3 to stage 4, and so on.

Let us look at how we can use this principle in a more general context, by considering the following problem of optimal allocation of resources, known as the **knapsack problem**. A person is faced with an overall weight limit 7 pounds, and has the choice of allocating this capacity between items of three types, A, B and C.

Each unit of item of type C weighs 1 pound, each unit of item of type B weighs 2 pounds and each unit of item of type A weighs 3 pounds. Obviously, there are many combinations posssible, such as 7 units of item C, or 1 unit of item A, and 4 of item C and so on. The criterion for making the best choice is that the person has made an estimate of the relative satisfaction he or she will get from taking these items along, and thinks that each unit of item A will give 8 units of satisfaction, each unit of item B 5 units, and each unit of item C 2 units, and the person wants to get the largest amount of satisfaction possible. (Sometimes a distinction is made that if only 1 unit of each item can be taken at the most, it is a **knapsack** problem, and if more than one unit of each item can be taken, it is a **cargo-loading** problem. However, the principle involved is the same).

The information about the items can be summarized in a table as below:

Type	Weight per unit	Satisfaction per unit
C	1	2
B	2	5
A	3	8

One could think of this as a multi-stage decision-making problem, the first stage being a decision on how many units of item C to take, the second stage how many units of item B, and the third stage how many units of item A. What about the 'change of state' in this problem? We can think of the person as starting with 7 units of spare capacity left, and the first stage decision, depending on whether the person decides to take 0, 1, or 2 , 3, 4 . . . 7 units of item C, leads to a new state of 7,6, 5, 4 . . . 0 units of capacity left. We could, as in the shortest route problem call the initial state start or rather start7, to indicate that there are 7 units of capacity available at that stage. Similarly one can call the new state when no units of C are chosen as one7 to indicate that we are now at the end of stage one, and that there are 7 units of capacity still available. Similarly, the new state when the decision is to take one unit of C would be one6, and so on for states one5, one4 . . . one0. We can show this as a network. Corresponding to the distances shown in the shortest route problem, we can show the benefits associated with each arc, as in Fig. 3.6.

Now in the second stage, from state one7 the possible decisions are to take 0, 1, 2 or 3 units of item B, leading to states we can call two7, two5, two3, and two1, with benefits respectively of 0, 5, 10 and 15 units, since each unit of item B gives 5 units of benefit. The names of the states have been chosen to give us some information about the capacity still available for use — if out of the capacity of 7 units available in state one7, 4 are used up for taking two units of item B, then at the end of the second stage, there will be 3 units of capacity left, leading to state two3. Again, from state one1 or state one0 the only possible decision is to take 0 units of B, since each unit of B requires capacity of 2 pounds. This will lead in the two cases to states two1 and two0 respectively. We could extend the network by showing the arcs and the benefits associated with that transition, for example the arc from one2 to two0 will be as shown in Fig. 3.7.

However, drawing even with these many states the diagram would be cumbersome, and impossible with larger numbers of states. Instead, the method of subtables we used in the shortest route problem would be handier. The tables for stages 1 and 2 would be as follows:

Table for stage 1

From	To							
	one7	one6	one5	one4	one3	one2	one1	one0
start7	0	2	4	6	8	10	12	14

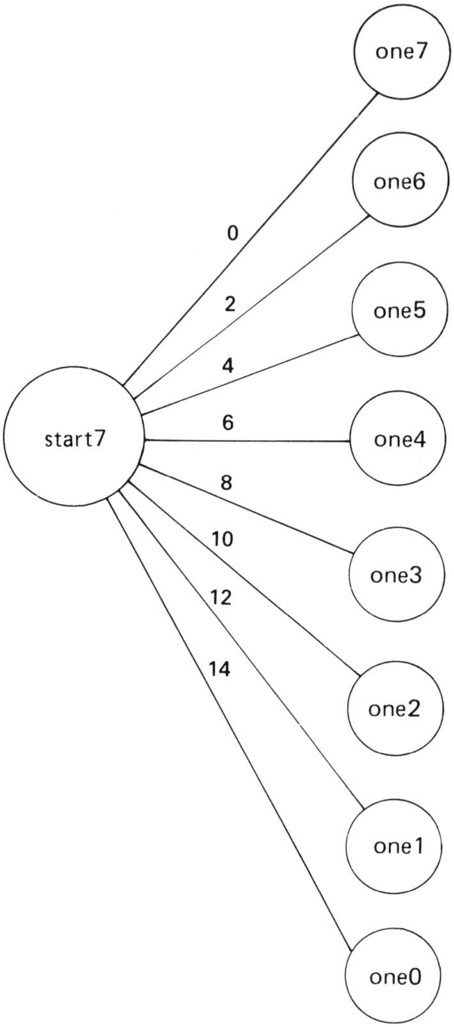
one7
one6
one5
one4
one3
one2
one1
one0
start7
0
2
4
6
8
10
12
14

Fig. 3.6.

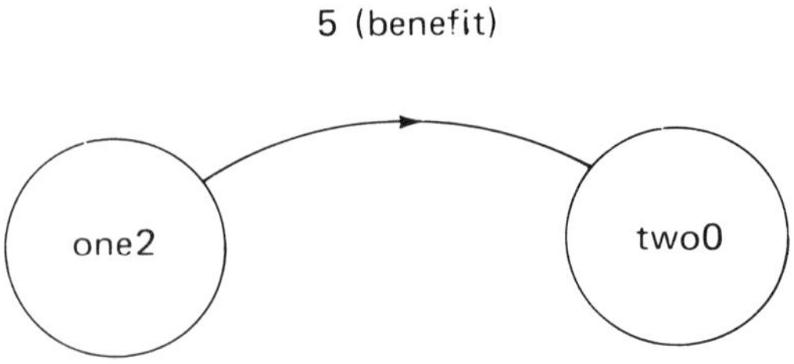
5 (benefit)
one2
two0

Fig. 3.7.

Table for stage 2

From	To							
	two7	two6	two5	two4	two3	two2	two1	two0
one7	0	x	5	x	10	x	15	x
one6	x	0	x	5	x	10	x	15
one5	x	x	0	x	5	x	10	x
one4	x	x	x	0	x	5	x	10
one3	x	x	x	x	0	x	5	x
one2	x	x	x	x	x	0	x	5
one1	x	x	x	x	x	x	0	x
one0	x	x	x	x	x	x	x	0

We see there are many impossible transitions, indicated as x (why is the transition from one3 to two2 impossible? Why are there entries only in alternate columns in rows one7 . . . one2 ?)

Finally, in the third stage, since the idea is to use up whatever capacity is available, we designate one final state , which we may call threedest, to show it is the destination at the end of stage three. The table would be:

Table for stage 3

From	To threedest
two7	16
two6	16
two5	8
two4	8
two3	8
two2	0
two1	0
two0	0

Notice that since in the third stage it makes sense to use up as much as possible of available capacity, if there are 6 or 7 units of capacity available, the best we can do is to take two units of item A for benefit of 16, since each unit of A requires 3 units of capacity. Similarly, if the available capacity is 3, 4 or 5, we can take one unit of A, and for less capacity we cannot take any unit of A.

We see the structure of the problem is the same as the shortest route problem — the only difference is that now we want the 'longest route' or maximum satisfaction. We start by considering the table for stage 3, and immediately read off the maximum capacity for the various states two7, two6, ... etc., since the only route from these states is direct to the destination, threedest, with benefits as given in the table.

Now, we can recurse to the table for stage 2 and apply the same kind of analysis as we did in the shortest route problem.

From the table for stage 2 given earlier, we see that from state one7 it is possible at stage2 to go to each of the following states, with corresponding benefits:

	two7	two5	two3	two1
Benefits	0	5	10	15
From table for stage 3 we already know that the best benefit from these same states would be	16	8	8	0

So we can conclude that the 'longest route' from one7 is the one through two3, giving a total of 10+8=18, as compared with the totals of 16, 13 and 15 for the routes through two7, two5 and two1 respectively. What this means is that if at stage 2 one has 7 units of capacity available, the best policy is to use up 4 units of capacity for a benefit of 10, and thereby go to state two3, and then use those 3 units of capacity for a benefit of 8, giving a total of 18. The benefits for the other states can be similarly analysed.

The reader may like to verify that by similar calculations the longest route from the other states are as given below:

From	one7	one6	one5	one4	one3	one2	one1	one0
longest route:	18	16	13	10	8	5	0	0
Now we can recurse to stage 1. We know from the Table for stage 1 that going from start to the above states would yield benefits of:	0	2	4	6	8	10	12	14
Total	18	18	17	16	16	15	12	14

So we can conclude that going from start to one7 or to one6 would be equally good and lead to a final total benefit of 18.

The route through one7 means 0 units of item C. From one7 we saw that the best route was through two3, i.e. taking 2 units of item B. From two3, the best decision was to take 1 unit of A. So, this route prescribes 0 units of C, 2 units of B and 1 unit of A, which uses up total capacity of 7, and gives benefit of 18.

The alternative route through one6 implies 1 unit of C. From one6, the best route was through two6, i.e. 0 units of B. And from two6, the best decision was to take 2 units of A. We can see that 1 unit of C plus 2 units of A would also give benefit of 18 and use capacity of 7.

Clearly, the calculations would become laborious for larger problems. But the program we have developed for finding the shortest route can be modified to find the largest route. What is needed is to change the 'smallest' clauses to 'largest' clauses, so that the longest route is picked instead of the shortest route. This is one of the exercises at the end of this chapter. A listing with this modification is given in the solutions section, with the name of the clause changed from sp for shortest path to knap1 for knapsack. In this case,we would invoke the program by asking:

```
?-knap1(From,threedest,Answer).
```

and the response would be:

```
From = two7
Answer = 16 ->;

From = two6
Answer = 16 ->;

From = two5
Answer = 8 ->;

From = two4
Answer = 8 ->;

From = two3
Answer = 8 ->;

From = two2
Answer = 0 ->;

From = two1
Answer = 0 ->;

From = two0
Answer = 0 ->;

From = start
Answer = 18 ->;

From = one7
Answer = 18 ->;

From = one6
Answer = 16 ->;

From = one5
Answer = 13 ->;

From = one4
Answer = 10 ->;

From = one3
Answer = 8 ->;

From = one2
Answer = 5 ->;

From = one1
Answer = 0 ->;

From = one0
Answer = 0 ->;
no
?-
```

We see that the response shows the final result and the intermediate results as well.

Let us however consider here one question: how would we represent the information in the tables above so that it conforms to the pattern we used for the shortest route problem? We would have to give it the clauses regarding sequence of the nodes and the 'distances' between the nodes, i.e. the benefits associated with the transitions, exactly as in the shortest route problem. For the example above, the required clauses would be:

```
/* CLAUSES TO BE PUT IN KNOWLEDGE BASE TO
REPRESENT THE INFORMATION IN THE TABLES FOR
DIFFERENT STAGES */
/* stage 1 from start */

follow(start,[one7,one6,one5,one4,one3,one2,one1,one0],
        [0,2,4,6,8,10,12,14]).

/* stage 2 starting from one . . . states */
follow(one7,[two7,two5,two3,two1],[0,5,10,15]).
follow(one6,[two6,two4,two2,two0],[0,5,10,15]).
follow(one5,[two5,two3,two1],[0,5,10]).
follow(one4,[two4,two2,two0],[0,5,10]).
follow(one3,[two3,two1],[0,5]).
follow(one2,[two2,two0],[0,5]).
follow(one1,[two1],[0]).
follow(one0,[two0],[0]).

/* stage 3 starting with two . . states ending threedest */

follow(two7,[threedest],[16]).
follow(two6,[threedest],[16]).
follow(two5,[threedest],[8]).
follow(two4,[threedest],[8]).
follow(two3,[threedest],[8]).
follow(two2,[threedest],[0]).
follow(two1,[threedest],[0]).
follow(two0,[threedest],[0]).
```

CONCLUSION

We have looked at a simple knapsack problem to highlight how a multistage decision process has a structure essentially the same as that of finding a path in a network, and how finding the best policy or sequence of decisions is equivalent to finding the shortest path. There are a variety of applications of this approach to situations which have essentially the same structure. For example, let us say a decision has to be taken on allocating a team of 6 salespeople to three different territories, A, B and C. The increase in sales from allocating 1, 2, 3 or 4 different workers to each territory are as follows (at least one worker is required in each territory so not more than 4 can be assigned to any one territory):

Number of workers	A	B	C
1	4	3	5
2	6	6	7
3	9	8	10
4	11	10	12

This could be considered as a multi-stage process, in which the first stage is the decision on how many to allocate to A. The table would be:

	To			
	one5	one4	one3	one2
Benefit	4	6	9	11

As in the previous example discussed, the stages one5, one4 ... represent states in which 5, 4, ... more workers are available for allocation.

At stage 2, transitions are possible from one5 to two4, two3 and two2, from one4 to two3 and two2 ... and so on.

At stage 3, from all the possible starting states two4, two3, two2 and two1, the transition would be to the destination or end state by using all the available persons (4, 3, 2 or 1), since the aim is to maximize sales. One could therefore solve this as a longest path problem exactly as in the previous cases. In Prolog notation, the information would have to be put in as:

```
follow(start,[one5,one4,one3,one2],[4,6,9,11]).
follow(one5,[two4,two3,two2,two1],[3,6,8,10]).
follow(one4,[two3,two2,two1],[3,6,8]).
follow(one3,[two2,two1],[3,6]).
follow(one2,[two1],[3]).
follow(two4,[dest],[12]).
follow(two3,[dest],[10]).
follow(two2,[dest],[7]).
follow(two1,[dest],[5]).
```

In the next chapter, we will extend the idea of finding paths to situations where there is some uncertainty as to the next node when one makes a transition from a given node.

EXERCISES

1. In the discussion of the sp program for finding the shortest path, we saw that when we asked a question like:

```
?-sp(bb,dest,Answer).
```

the program would find the immediate successors of bb, i.e. [dd,ee] and then go to the 'best' clauses. When using these clauses, it would recurse using each of the immediate successors, first solving sp(dd,dest,Answer) and then sp(ee,dest,Answer). The program also inserts these two answers into a list Choices, so that later it can find the smallest of the distances. If we want to modify the program so that it will give the actual path as well as the shortest distance:

(1) First, the clauses for 'best' would have to be modified so that when it recurses, it carries along not only the name of the node (dd or ee) but the shortest path found from that node to the destination.
(2) Second, after all the immediate successor nodes have been considered, the program goes to the 'smallest' clauses. It is carrying along a list of numbers, showing the shortest paths through each of the immediate successor nodes, and finds the smallest. As we saw in the discussion in the body of the chapter, it carries along a list [9,11] and finds the smallest number. But now it is carrying along a **list of lists**, in which each list is one of the immediate successor nodes plus the path from that node to the destination. This means that the smallest clauses will have to be modified

so that they select not just the smallest number, but the list headed by the smallest number.

(3) Third, when these changes are made to the 'best' and 'smallest' programs, one will have to think of the consequent changes required for the main 'sp' clauses themselves.

The reader might like to think about the problem on the basis of these guidelines and try to make the necessary modification. The solutions section discusses the modifications in the light of these guidelines.

2. Having developed the program in 1 above, we might want to modify it so that we can have a program which will find the shortest route from start to any node, instead of shortest route from any node to destination. The following considerations are relevant:

(1) it would be convenient to represent the relationship between nodes in a slightly different form. We might want to show for each node what its list of immediate predecessors is, instead of what its immediate successors are, because now we are going to recurse from 'front to back' whereas previously we went back to front. For instance, for the same problem, the first clause in the knowledge base could be:

```
before([start],aa,[5]).
```

This would convey that start is the list of predecessors of aa. Another clause would be: before([aa,bb],dd,[4,6])., and so on.

(2) Since we are now operating with all predecessors of a given node, rather than all successors, it would be more informative to use a variable Allpred in the sp and best clauses instead of Allsucc.

A modified listing, and the output produced by invoking the program, are included in the solutions section.

3. We saw in the chapter that we need to change the shortest route program into a longest route program for solving the knapsack problem. The main requirement is that the 'smallest' program given below should be suitably changed to make it the 'largest' program:

```
/* SMALLEST CLAUSE 1 */

smallest([Only],Only).

/* SMALLEST CLAUSE 2 */

smallest([H1,H2|T],Result):-
        H1>=H2,
        smallest(H2|T],Result).

/* SMALLEST CLAUSE 3 */

smallest([H1,H2|T],Result):-
    H1<H2,
    smallest([H1|T],Result).
```

Consider the working of the program as discussed in the chapter, and make the necessary modification in clauses 2 and 3.

4

Transition networks and uncertainty

PREVIEW

There are many examples of systems which change from an initial state to one of a set of possible states. This chapter looks at the use of transition networks in modelling this kind of situation.

INTRODUCTION

We have seen in the last three chapters that the concept of a transition network is general enough to be applied in a wide variety of contexts. However, there was one feature in all the examples we have seen which may not be there in certain situations. In all of those examples, there was no uncertainty at all as to which state or node would be reached if the decision was taken at the current node to choose a particular arc or link leading away from that node. One chooses a particular road, and one knows for certain which city it leads to. If one chooses to pour from jug_a to jug_b, one knows for certain what the next state of the system would be. If we decide to put a certain number of units of item A into the knapsack, we are sure how much capacity will be left, and so on. This characteristic of being certain of the outcome is described by saying that the transitions are **deterministic**.

In contrast with this, there are situations where the result of a transition from a current state is not one specific state but one of a set of possible states. For example, if one takes a bet, then the result could be that one wins or loses, so one could say that there is a set of (two) possible states, 'win' or 'lose'.

The major conceptual tool which has been developed over the last few centuries for dealing with such non-deterministic situations is the theory of probability, and in the next section we will review the fundamentals of this theory and see how transition networks are used in applications of probability theory.

BASICS OF PROBABILITY THEORY

The classic example is of course tossing a coin, where we know the next state of the coin is either heads or tails. One cannot carry the analysis much further if all that one knows is that the next state is uncertain. But in many cases, one is also able to estimate some kind of ratio for the relative degree of uncertainty of the different outcomes. In the case of the coin, one finds it plausible to say that the odds are 1 to 1 of heads or tails, or, to put it slightly differently, that they are equally likely and the ratio of their uncertainties is 1. A useful way of looking at the odds is to think of it as follows. In a deterministic situation, we are fully confident of the outcome. In a non-deterministic situation, we partition or 'share out'our confidence between different outcomes. In the case of the coin, we are spreading out our confidence in the ratio 1 to 1. Similarly, if we say that the odds are 1 to 6 of a certain team winning the championship, we are sharing out our confidence in that ratio between the possibilities of the team winning and losing. This leads to the other concept associated with such situations, the concept of probability. If the odds are 1 to 1, we could say that we are distributing our confidence 1/2 and 1/2, or say that the probability of each outcome is 1/2. Similarly, if the odds are 1 to 6, we are distributing our confidence 1/7 to one outcome and 6/7 to the other outcome and say the probability of winning is 1/7 and of losing 6/7. If there are three outcomes, we might in a particular case think of the odds or ratios of the uncertainties as 2 to 2 to 3, or it terms of probability, that the probabilities are 2/7, 2/7 and 3/7. In other words, the probability of a particular outcome is the proportion of our confidence or belief that we attach to that outcome. Looking at it this way, we see why the first basic rule of probabilities is that the the probabilities of the complete set of non-overlapping possibilities must add up to 1. The phrase **mutually exclusive and collectively exhaustive** is usually used to indicate that only one of the various possibilities can materialize and that all possibilities have been taken into account.

Now, where do we get the odds or probabilities? There are broadly speaking three different ways in which the odds are estimated. Firstly, from the general nature of the situation one may be able to say that certain odds or probabilities seem plausible or natural. This is the case for tossing a coin, where on the face of it it seems that if the coin is not weighted to one side or the other, it is equally likely to come up heads or tails. Secondly, there may be some situations where there is some statistical basis for the odds. For example, when the weather forecast says the probability of rain tomorrow is 8/10 or 0.8, it is based on the fact that over a reasonable period of time, conditions similar to those prevailing today were followed by rain in 8/10 of the cases, and in the other 2/10 of the cases there was no rain. Such probabilities are based on *relative frequencies* of occurrences of the event whose probability is estimated. Another example of a probability based on relative frequency would be this: if according to the census, 20% of people in a particular country have a university degree, then we can say that if we select some person at random, the probability that he or she has a university degree is 1/5 or 0.2. (There is one point one has to be careful about, the concept of randomness. Obviously, if one selects a person in some university town, the chances of finding a person with a university degree would be considerably higher, or if one goes to an area where the underprivileged live, the chances would be lower. The concept of randomness involves the idea that the sample we draw should have an equal chance in relation to any other possible sample which could be drawn.) Finally, there is a third category of situations in which there is no such relative frequency or statistical basis for the estimate of probabilities, nor is one able to give any *a priori* plausible estimate of the probabilities. For instance, if someone says that there is a 80% chance, or probability of 0.8, that the political union of Europe will increase everybody's happiness, and another person says he thinks there is only a 0.3 probability of that happening, there is no clear statistical basis on which these beliefs are based. Similarly, if someone says she thinks the odds of team A, team B or team C respectively winning the championship are 2 to 3 to 4, in effect she is saying that B is 1.5 times as likely to win as A, and C is twice as likely to win as A. How does one find an underpinning for these ratios of 1.5 and 2? One should regard these estimates as expressions of **subjective belief**. It might seem on the face of it that such subjective estimates of probabilities may not have much value, but one of the things we will look at in this chapter is how such subjective estimates can provide a starting point for further analyses. These analyses would allow one to suitably revise one's initial beliefs in the light of additional information. We can conveniently represent probabilities in the form of a transition network. For example, for the coins case, we can say that from 'start', there is a transition to the state 'heads' or to the state 'tails'. Just as we associated a distance or a profit or payoff with each branch of a network in previous examples, we can associate the probability with each arc of the network, as in Figs 4.1 and 4.2 and so on. If there are more than two possibilities, naturally we would draw as many branches from start as there are possibilities.

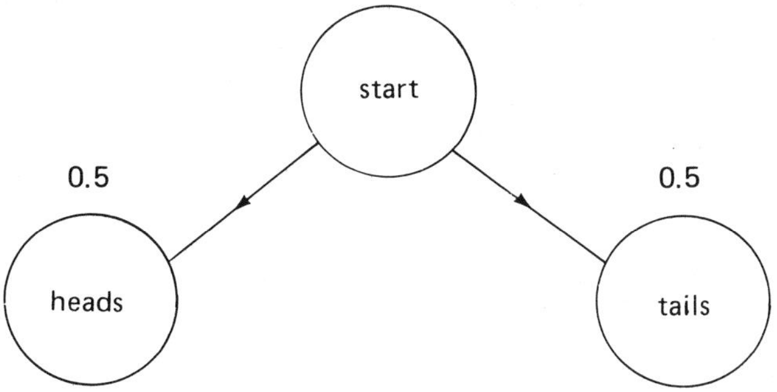

Fig. 4.1.

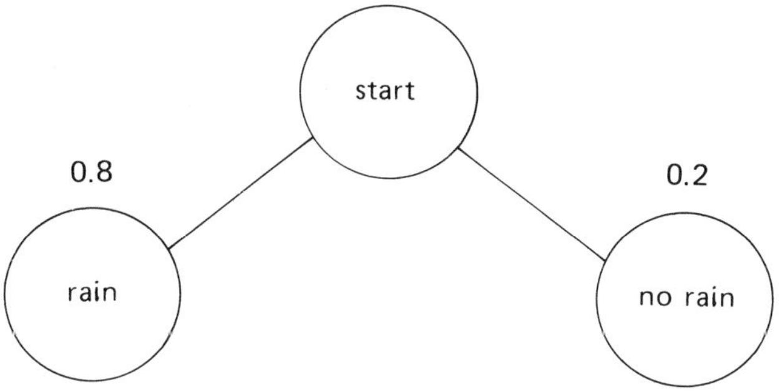

Fig. 4.2.

SEQUENCES OF EVENTS

The next important aspect of the theory of probabilities relates to the probability of a sequence of non-deterministic occurrences. If we know the probability of each of the occurrences in a sequence of occurrences, how do we combine these probabilities to find the probability of the whole sequence?

Let us say the forecast on Friday for weather on the weekend is that there is 0.8 probability of rain on Saturday. For Sunday, the forecast is that *if* it rains on Saturday, the probability of its raining on Sunday is only 0.3, but if it does not rain on Saturday, the probability of rain on Sunday is 0.7. To find out the probability that it will rain both on Saturday and on Sunday, we can conveniently apply the concept of sharing out our confidence that we considered earlier. We can draw first a diagram showing that on Saturday, we are attaching probabilities of 0.8 and 0.2 to rain and no rain (Fig. 4.3). The

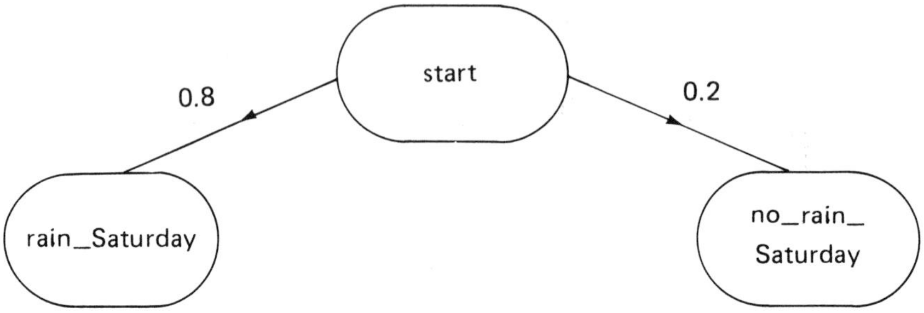

Fig. 4.3.

forecast that if it rains on Saturday, the probability of rain on Sunday is 0.3 means that we are further sharing out the 0.8 certainty we have allotted to Saturday in proportions 0.3 and 0.7 to rain or no rain on Sunday. So the left half of our diagram goes down one level (Fig. 4.4). From this we can see that

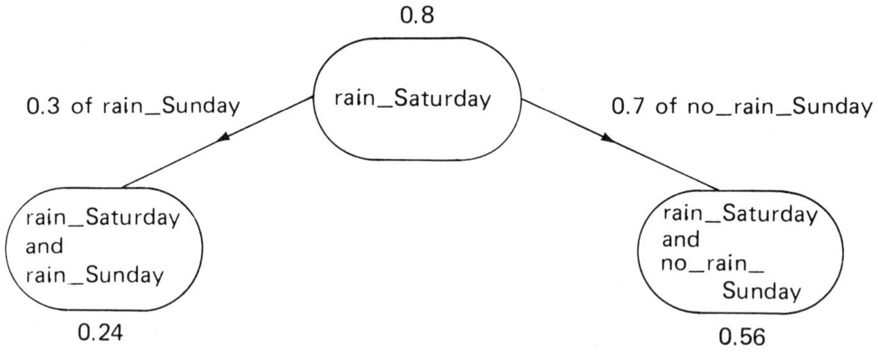

Fig. 4.4.

since only 0.3 of our certainty of rain on Saturday is allocated to rain on Sunday, and our certainty of rain on Saturday is itself only 0.8, the probability of rain on Saturday and on Sunday is $0.8 \times 0.3 = 0.24$. In the same way, the probability of rain on Saturday but not on Sunday is $0.8 \times 0.7 = 0.56$. Another way of looking at it is to say that our certainty of 0.8 of rain on

Saturday has been partioned into 0.24 and 0.56, in the ratio 3 to 7, which are the odds of rain or no rain on Sunday *if* it rains on Saturday.

By the same line of reasoning, we can partition the right half of Fig. 4.3, 0.2 probability of no rain on Saturday, into 0.14 probability of no rain on Saturday and rain on Sunday, and 0.06 probability of no rain on Saturday, or on Sunday (Fig. 4.5). Formally, this is expressed in the theory of probability

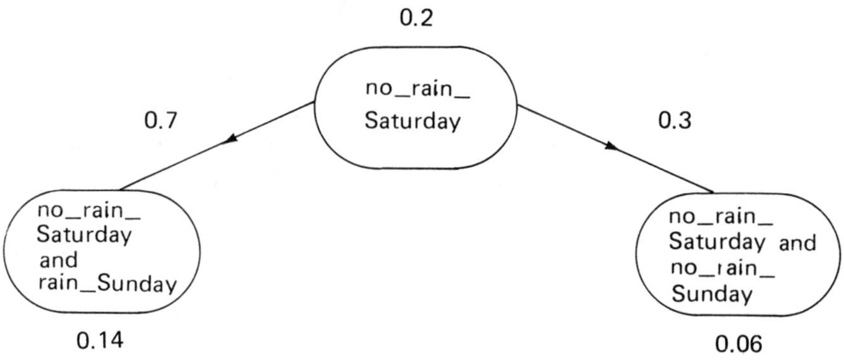

Fig. 4.5.

by saying that the joint probability of both the events (say, rain on both days) is the probability of one event (rain on Saturday, 0.8), multiplied by the *conditional probability* of the second event (rain on Sunday), given that the first event occurs, i.e the probability of the second event if the first event has occcurred (in this case 0.3). Each of the four branches in Figs 4.4 and 4.5 is the joint probability of different pairs of events, and naturally the joint probabilities of all the possible pairs of events also add up to 1, since they are the final sharing out of the total certainty of 1.

A special case which might arise is when the probability of the second event is independent of the first event. In the example above, the probability of rain on Sunday is 0.3 if it rains on Saturday, and is 0.7 if it does not rain on Saturday, i.e. its probability is different, depending on whether it was rain or no rain on Saturday, so it is not independent of the first event. What if it is independent, i.e. it is the same whatever the outcome of the first event? All we need to do is to use this single probability on both branches of the diagram. In more general terms, if probability of event 2 given outcome 1 for event 1 is the same as probability of event 2 given outcome 2 for event 1 (and so on, if there are more possible outcomes for event 1), then we might as well just call it the **unconditional probability** of event 2.

Clearly, we can extend this to a sequence of 3, 4, 5, . . . events, and say that the joint probability of event 1 and event 2 and event 3 is the probability of event 1 times conditional probability of event 2 given event 1 times conditional probability of event 3 given event 1 and event 2 . . . and so on. In the case where the events are independent, the conditional probabilities are not really conditional but unconditional, so we just multiply the unconditional probabilities together. An example is the tossing of a coin 2, 3, 4 . . . times. With a fair coin, the probability of a head on each flip of the coin is 0.5, no matter what happened on previous tosses, so the probability of two heads in a row is 0.5×0.5, of three heads in a row $0.5 \times 0.5 \times 0.5$, and so on.

Looking at the diagram in the example above also enables us to answer further questions. If we want to know what the total probability of rain on Sunday is, we see that two of the four branches involve rain on Sunday, and adding these up gives us the total probability of rain on Sunday as $0.24+0.14=0.38$. It follows that the probability of no rain on Sunday must be $1-0.38=0.62$, a result we could have also found from adding up the probabilities on the two branches with no rain on Sunday.

We could add up the 0.24 and 0.14 to get 0.38 and the 0.56 and the 0.06 to get 0.62 because these were joint events which were non-overlapping or mutually exclusive, i.e. one or the other could occur, but not both. What about events which are not mutually exclusive, i.e. sometimes one occurs but not the other, but sometimes both occur together? For instance, if we ask: What is the probability that it will not rain on at least one of the two days? We know at the beginning that the probability of no rain on Saturday is 0.2 and we have found the probability of no rain on Sunday to be 0.62. But it would be incorrect to add these up and say that the probability it will not rain on at least one of the days is 0.82. We can see that it is incorrect by considering the following. The probability that it does not rain on at least one of the days must be 1 minus the probability that it rains on both days. But the probability that it rains on both days is from our diagram 0.24, so this line of reasoning tells us the probability that it will not rain on at least one of the days is 0.76 and not 0.82.

The problem is that when we added up up the probability of no rain on Saturday and the probability of no rain on Sunday, we had considered two things which were *not* mutually exclusive, and as a result have double counted something, namely the probability of no rain on both of the days. The probability 0.06 of no rain on both the days is part of both the 0.62 and of the 0.2. So if we subtract this from the 0.82 we get the correct answer of 0.76, which is a component of both these events.

This conclusion is usually formally expressed in the form that the probability of at least one of two overlapping events is the sum of their probabilities minus the probability of both of them occurring. Actually, it is simpler to think always in terms of non-overlapping events and say that the probability of at least one of two overlapping events consists of three non-overlapping events:

(1) the first but not the second
(2) the second but not the first
(3) both together.

Looking at it this way, we would have seen in the above example that (1) is 0.56 (2) is 0.14 and (3) is 0.06, which add up to 0.76. A helpful way of looking at this is in terms of what is called a Venn diagram, eponymously named for the nineteenth-century mathematician who first developed them. This technique consists of drawing a box to show the complete set of all possibilities, representing 1, and drawing circles within the box to represent the different events we are considering, as in the Fig. 4.6. We can see in this diagram that

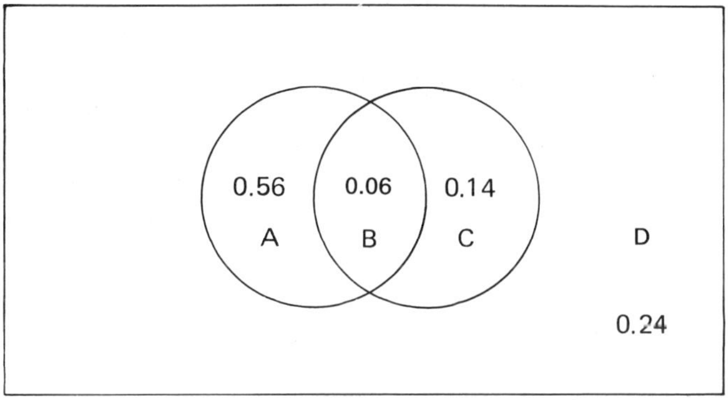

Fig. 4.6.

the areas A, B and C represent (1), (2) and (3) in our analysis above. In addition area D represents the only possibility left, i.e. no rain on both of the days.

In our discussion above, we have thought of a transition as being a deterministic or non-deterministic process. It is often useful to think of the outcome of the transition as being a deterministic or non-deterministic **variable**. When we toss a coin, we can think of the result as a variable, 'face which is upward', which has one of two possible **values**, 'heads' or 'tails'. Similarly, we can think of the weather on Sunday as a variable, which can have one of two values, rainy or not rainy. We can say the outcomes in these examples is a probabilistic variable. An alternative name is to call it a **stochastic** variable, a name derived from a Greek root for guess or conjecture. Also, typically, the outcome can be associated in a meaningful way with a number. Again, going back to the coin tossing example, we can say that the outcome is either 1 head or 0 heads. We could also have defined it as either 0 tails or 1 tail. In any case, when the outcome is associated with a number, the number is also called a **random variable**. But since this use of the word random is somewhat different from the usual meaning of everything having an equal chance (as when we, say, pick one book at random), it would be preferable to use the name stochastic variable even when there is a number associated with the outcome. One other basic concept of importance in regard to probabilistic processes is that of **expected value** of a stochastic variable. This concept is applicable only in cases where the outcomes are expressed as numbers rather than in terms of qualitative descriptions. The expected value of a stochastic variable is the long-run average value of the variable, that is the average value one might expect if the process is repeated over and over again. For example, when we toss a coin once, we expect either 1 head or 0 heads. But if we keep flipping it again and again, 100, 1000, 10 000, . . . times, intuitively we expect that half the time we would get 1 head, and half the time a tails or 0 heads. So, if say in 10 000 repetitions we get 5000 heads and 5000 tails, the average number of heads would be the total number of heads, $(5000\times1)+(5000\times0)$, divided by 10 000, which is the same as $(0.5\times1)+(0.5\times0)=0.5$. So, the expected number of heads in one toss of a coin is 0.5. We can generalize this and say that if a stochastic variable has possible values V1, V2, V3 . . . with associated probabilities p1, p2, p3 . . . the expected value would be $(p1\times V1)+(p2\times V2)$ . . . . This is sometimes called the sum of the values **weighted** by the respective probabilities. We can see that when we toss a coin, we get 1 head or 0 heads, so 0.5 is not really the number of heads we 'expect' to get, in the common use of the word expect, but only in the technical sense in which 'expected value' has been defined. Similarly, if we think there is a 1/5 or 0.2 chance of winning \$300 in a particular bet, and 4/5 chance of losing \$100, we say that the expected value of our gain is 0.2 times 300 plus 0.8 times $(-100)=-20$. The important thing to keep in mind in reference to the expected value of a probabilistic variable is that it is based on the idea of what would happen in a very large number of repetitions of the process. If we take the bet a small number of times, say twice, the expected value of -20 does not directly tell us what is going to happen. Either we win both times, in which case we gain \$600, or lose both times for a total loss of \$400, or win once and lose once, for a gain of \$200. How do these figures, gain of \$600, or loss of \$400 or gain of \$200 in two repetitions connect up with the expected value of the bet? Let us think of betting twice as itself a process which can be repeated a large number of times, and the outcome as a stochastic variable which can have three possible values, 600, -400, or 200. Now what are the probabilities of getting these values?

Occurrence	Value of variable	Probability
Winning twice	600	0.2×0.2=0.04
Losing twice	−200	0.8×0.8=0.64
Win first time lose second time	200	0.2×0.8=0.16
Lose first time win second time	200	0.8×0.2= 0.16

The expected value of the variable is therefore got by finding the weighted sum of the values, using the probabilities as weights:

$$(0.04\times600)+(0.64\times-200)+(0.32\times200)=-40\ .$$

Since the expected value or average gain in each repetition is −20, the average gain in two repetitions comes out as −40. Even though it is only in a long series of replications that the expected value becomes meaningful, the concept does have a useful role to play in decision-making, as we will see below.

DECISION TREES

Let us consider the following problem in which the situation is as follows. A decision maker has to choose between investing in three different projects, A, B, and C. The payoff for each of these projects is stochastic, and not deterministic and depends on whether the economy holds even or deteriorates. The net return from the projects is likely to be different in these different circumstances. The decision maker estimates the returns as likely to be as follows:

	Returns	
Project	If economy holds even	If economy deteriorates
A	20	15
B	35	5
C	30	10

But for the uncertainty associated with the future of the economy, the decision would be simple. If one knows the economy will hold even, obviously the best choice is project B. On the other hand if one knows the economy is going to deteriorate, the best choice is A. But because one does not know which of these uncertain outcomes is going to actually prevail, some technique for handling the uncertainty becomes necessary.

The worst case of uncertainty is when one has no kind of estimate of the relative chances of the different possibilities. In such circumstances, various criteria for choice have been suggested. One criterion is the optimistic one of hoping for the best possible payoff. That approach would lead one to choose B, since it has the highest payoff in the whole table. But obviously if one's optimism is misplaced and the economy does not hold good, one would have been better off choosing A or even C.

A second criterion is the pessimistic one of thinking that one should compare the worst result of each choice, and then pick the project for which this worst result is the highest of the three worst results. Using that approach, one would look at the worst possible results of choosing A, B, C respectively, i.e. 15, 5 and 10, and choose A. Of course, if the economy does hold steady, one would regret that one gets only 20 whereas choosing B would have given a payoff of 35. This criterion is also called the **maximin** approach, for it is based as we saw on looking at the worst case, or minimum, that each choice would lead to, and then picking the best or maximum of these minima.

Suppose, however, that one has some kind of rough estimate of the probabilities of whether the economy will hold steady or worsen. Suppose the decision maker estimates these probabilities to be 0.6 and 0.4. (We will come back later as to how these probabilities might be estimated, and how even subjective guesses of these probabilities could be used as starting points for further analysis.)

An approach which has been suggested is to compare the expected value of the return for each of the choices, and pick the one with the best expected value. As we saw earlier, the expected value is the weighted sum of the possible values, and so for the stochastic variable return from project A, the expected value is $(0.6\times20)+(0.4\times15)=18$.

Similarly, for projects B and C, the expected values of the returns are:

$$(0.6\times35)+(0.4\times\ 5)=23\ .$$

$$(0.6\times30)+(0.4\times10)=22\ .$$

So, using this criterion, project B has the highest expected value, and should be selected.

A conceptual problem with using the expected values to compare such one-shot decisions arises from the very nature or definition of the expected value of a stochastic variable. As we saw, it tells us what the long-run average value is likely to be in a long series of repetitions. But in this decision, we are considering a one-shot operation, and we know the result of choosing project B is going to be either 35 or 5, and the expected value does not have anything directly to say about this. Nevertheless, it is used as a criterion just as a rough index for ranking the choices in some relative order.

Of course, there may be situations where in fact there are a large number of repetitions of the same choice, and in that case the use of the expected value as the criterion for comparing decisions can be justified much more strongly.

We can extend our previous representation in terms of a graph or network to incorporate the idea of expected values as well. We can end each branch with the value resulting from that branch, and then attach the expected value to the node from which the branches start. For example, we can represent the relevant facts about project A as is Fig. 4.7. Similar diagrams can be drawn

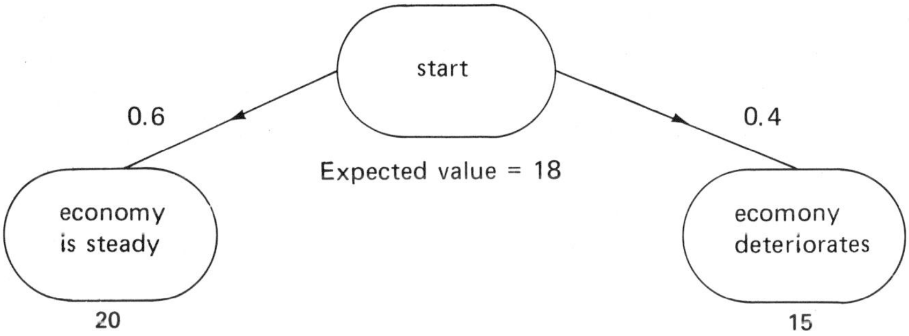

Fig. 4.7.

for projects B and C. Now we can combine all these into one network (Fig. 4.8).

In this representation in the form of what is called a **decision tree**, we have to distinguish between two different types of nodes. The start node is a **choice** node, and one goes by only one of the branches, depending on the decision taken. On the other hand, the nodes for A and B are probabilistic nodes, where there is a resulting expected value depending on the probabilities. The process of making a decision using this form of representation is sometimes referred to as 'rolling back'. One evaluates the probabilistic nodes to get expected values, and then chooses one node from the choice node to make the best decision. We will look a little later at a more involved example which includes more choices nodes and decision nodes, and not just one of each.

Decision-making using the expected value of payoff can be considered

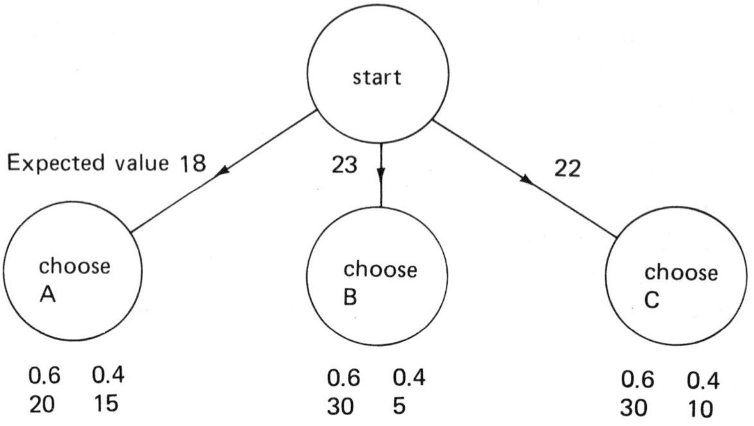

Fig. 4.8.

from another angle. The essence of uncertainty is that one may have to take a decision which might not be the one which one would take if one had full information. This aspect of decision-making under uncertainty has been captured by introducing the idea of **regret** into the analysis. The idea is that the amount of regret would be 0 for the particular decision which one would take if one had full information, and there would be some regret associated with each of the other decisions. Applying this to the example above, the best decision if one knows that the economy will hold steady is to choose project B, so it is assigned a regret of 0. Projects A and C are assigned regrets of $35-20=15$ and $35-30=5$ respectively, since that is the extent to which they will be later seen to have fallen short of the payoff of 35 associated with B, if one selects those projects and it turns out later that the economy does hold steady, and one realizes that B would have been the best choice. Similarly, if one knew that the economy was going to deteriorate, the 0 regret decision is A, and B and C have respectively the potential for regret of 10 and 5. We can therefore assemble these figures into a table of regrets:

Project	Regrets	
	If economy holds even	If economy deteriorates
A	15	0
B	0	10
C	5	5

Looking at the regrets table, the pessimist would say that A and B have potential for maximum regret of 15 and 10, whereas C can at the most lead to regret of 5, and so C would be the best choice. For the optimist, the regret table is not meaningful, because his focus is on the actual payoff. One could also calculate the expected value of regret for each of the projects, and taking the weighted sums of the regrets for each project, we get these expected values as:

$$0.6\times 15+0.4\times 0=9 \text{ for A.}$$

$$0.6\times 0+0.4\times 10=4 \text{ for B}$$

$$0.6\times 5+0.4\times 5=5 \text{ for C.}$$

The best choice now would be the project which gives the least expected value of regret and this is B. B is also what we would choose by using the expected value of the payoff, and the fact that that choosing on the basis of the largest expected value of payoff and of least expected value of regret have led to the same choice is no accident, but has an inherent logical connection. Looking back at our calculation, we see that for the three projects the values are as follows:

	Expected value of	
	Payoff	Regret
A	18	9
B	23	4
C	22	5

The interesting thing about this table is firstly that the higher the expected value of the payoff, the lower the expected value of the regret, which is intuitively reasonable, for the more the long-run average gain, the less the long-run average regret. But there is something more in this table, that the expected values of the regret and the payoff add up in all cases to exactly the same number, 27. What does this signify?

We see that the expected value of gain falls short of 27 by the expected value of regret. So, if the expected value of regret were to be 0, then the expected value of gain would become 27. Now in what circumstances would the regret be 0? Let us imagine we have some perfect forecasting device, which in the various 'repetitions' of the decision, would tell us each time whether the economy is going to remain steady or worsen. Clearly, this perfect forecaster would tell us 0.6 of the time that the economy is going to remain steady, and knowing this we would choose B and get a gain of 35. The other 0.4 of the time, when the forecast is of deterioration of the economy, we would choose A and get 15 as gain. So, in those circumstances our long-run average or expected value of gain would be $(0.6\times35)+(0.4\times15)=27$, and our expected value of regret would be 0. This value, of what would be obtained with a perfect forecasting device available, is known as the **expected value with perfect information**. Of course, what would be available are only imperfect forecasts, liable to error. Nevertheless, the concept of the expected value with perfect information is useful — it gives us the upper limit of what can be achieved. To see how this knowledge of the upper limit is of value, let us extend the analysis further to consider what happens when we have a forecasting mechanism which is not perfect. However, we have some information about its reliability in the following form. The previous record of the forecasting firm shows that in 80% of the cases in which the economy remained steady over the next year, they had correctly forecast this, but 20% of the time they mistakenly forecast a downturn. On the other hand, 90% of downturns were correctly forecast by them, but 10% of the time their forecast was of a downturn, but the economy remained steady. Knowing this, what would be the best action to take if they forecast that the economy would remain steady? If they were perfect forecasters, we would of course choose project B as the best bet when they predict that the economy is likely to hold steady. But knowing their less than perfect record, how should we use their forecast?

REVISION OF PROBABILITIES USING BAYES' THEOREM

The nineteenth-century British clergyman the Rev. Thomas Bayes developed a way of analysing such situations which is known in his honour as Bayes' Theorem. The essence of Bayes Theorem is this: when we have some initial or prior estimates of probabilities (in this case 0.6 and 0.40), and we have some additional information which is not perfect but about which we have relevant information about how often it is right or wrong (in this case the forecasting record of the forecasters), then how do we use this additional imperfect information to revise or update our initial estimates? A convenient way of looking at Bayes' Theorem is in terms of the probability trees we have looked at earlier. We can draw a diagram as in Fig. 4.9.

Now, in the light of previous discussions of joint probabilities, we can see that there are four possible outcomes:

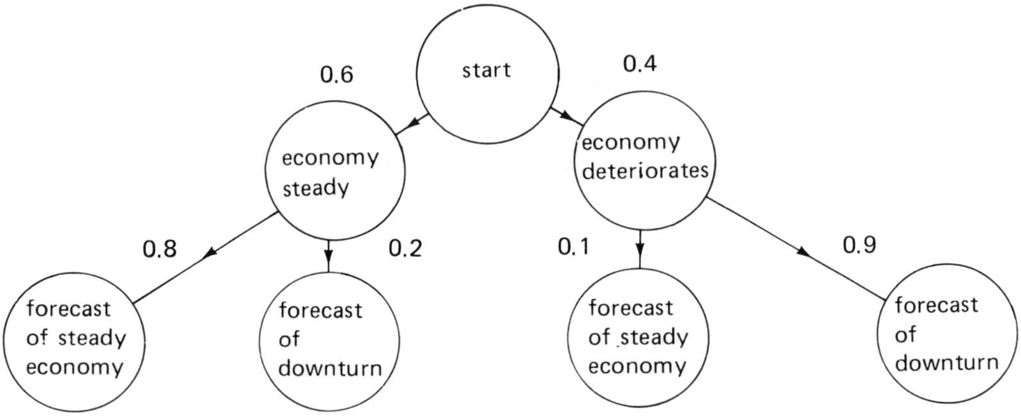

Fig. 4.9.

	Outcome	Probability
(1)	Forecast of steady economy and economy does remain steady	0.6×0.8=0.48
(2)	Forecast of downturn but turns out wrong	0.6×0.2=0.12
(3)	Forecast of downturn turns out correct	0.4×0.9=0.36
(4)	Forecast of steady economy but turns out wrong	0.4×0.1=0.04

We can see from this that the forecast is of a steady economy in 0.48+0.04=0.52 of the cases. But the proportion of these cases in which the economy does remain steady is 0.48/0.52 and the proportion of cases in which the economy deteriorates is the remaining 0.04/0.52.So when the forecast is of a steady economy, the odds are 48 to 4 that in fact it will remain steady.

We have revised our odds from our original 6 to 4 to this new value, on the basis of the additional information. Now, using these revised odds and probabilities, we can calculate the expected value for the three projects as:

$$A:\ (48/52)\times 20+(4/52)\times 15=1020/52$$

$$B:\ (48/52)\times 35+(4/52)\times 5=1700/52$$

$$C:\ (48/52)\times 30+(4/52)\times 10=1540/52$$

The best choice is still B, if the forecast is that the economy will remain steady.

What if the forecast is of a downturn? On the same lines, we can see from the diagram that the forecast of a downturn is made by the forecasters in 0.48

of the cases, and in 0.36 it is correct, and in the remaining 0.12 it is wrong. So, IF the forecast is of a downturn, we would revise our odds as 12 to 36, or 1 to 3, that the economy will remain steady or deteriorate, or we can say our revised probabilities are now 1/4 and 3/4. We can similarly calculate the expected values using these revised probabilities:

$$\text{A:}\quad (0.25\times20)+(0.75\times15)=16.25$$

$$\text{B:}\quad (0.25\times35)+(0.75\times\ 5)=12.5$$

$$\text{C:}\quad (0.25\times30)+(0.75\times10)=15.0$$

So, if the forecast is of a deterioration in the economy, it would be better to choose A.

How can we assess whether following the recommendations of the forecast and choosing A or B is better than choosing B all the time as we would have done in the absence of the forecasts? We know that the forecast is of a steady economy 0.52 of the time and of a downturn 0.48 of the time. So, when following the forecasts, our expected value would be:

$$\underset{\text{by choosing B}}{0.52\times1700/52)} \quad + \quad \underset{\text{by choosing A}}{0.48\times16.25=24.8}$$

So we see that the forecasts would enable us to increase our long-term average from the 23 we would get by choosing B all the time, though not up to the level of 27 which a perfect forecasting mechanism would achieve.

We can also verify from the more detailed analysis we have made that if we choose B all the time our expected value would be:

$$0.52\times(1700/52)+0.48\ \times12.5=23.$$

By looking at this, we see that what the forecast does is to enable us to take a decision more discriminatingly, rather than stick to B all the time.

It is not necessarily true that an imperfect forecast will always lead to an improvement, sometimes it may lead to zero improvement. To conclude this chapter, let us consider how we can design specifications for the kind of analyses we have considered above, and program them in Prolog.

In designing programs, it is useful to think in terms of the various modules and then fit them together. The first module we want is to calculate the expected value for each project. Representing information in the form of lists is handy here. We can have a list of returns, and a list of corresponding probabilities for each project. For example, we might put the relevant information on three projects aa, bb and cc into the knowledge base in the following form:

```
project(aa,[20,10,30],[0.3,0.5,0.2]).
project(bb,[15,30,10],[0.3,0.5,0.2]).
project(cc, [10,10,40],[0.3,0.5,0.2]).
```

Looking at project aa, what we would like is to ask a question like:

```
?-expvalue(aa,What).
```

and get back an answer.

To do this, we need to have something which will use the information we have supplied, i.e. the name aa, to search the 'project' clauses in the knowledge base, and pull out the list of returns, [20,10,30], and the corresponding probabilities, [0.3,0.5,0.2].

Having pulled these out, it needs first to multiply the corresponding returns and probabilities, $20\times0.3=6, 10\times0.5=5$, and $30\times0.2=6$, and then add these up. We had already seen in the appendix to Chapter 1 that if we have a list of numbers, we can add them up to get a result, so if we form a list of the three values [6,5,6], we can use the previous 'addup' clause to find the sum 17. The sum of the product of two series of numbers of equal length is called the **vector product** of the two series, so we will use that as the predicate name:

```
vecprod(List1,List2,Answer) :-
        vecproda(List1,List2,Result),
        addup(Result,Answer).

addup([],0).
addup([H|T],Result) :-
        addup(T,Rest),
        Result is H+Rest.
```

So, we need another program 'vecproda' which will produce the list [6,5,6] as Result, and we can then get Answer by using the 'addup' clauses.

We can generate the list bylooking at the problem in terms of recursion: The list of required values has:

as the head, the product of the heads of the two lists.
as tail, the vecproda of the tails of the two lists.

Also, as a boundary case for the recursion, when the two lists are empty, the result has also to be an empty list. Based on these considerations, we can design vecproda as follows:

```
vecproda([H1|T1],[H2|T2],[A|B]):-
                    A is H1*H2,
                  vecproda(T1,T2,B).

vecproda([],[],[]).
```

Finally, we need to put all these together in the main clause we will invoke:

```
expval(Project,Answer) :-
          project(Project,Returns,Probs),
            vecprod(Returns,Probs,Answer).
```

The output one gets by invoking this program in various ways is as follows:

```
?-expval(aa,What).

What=17.0 ->
yes

?-expval(WhichProject,What).

WhichProject=aa
What=17.0 ->;

 WhichProject=bb
What=21.5 ->;

WhichProject=cc
What=16.0 ->;

no
```

Now that we have the expval clauses which, for each project, give us the expected value, the next step is to think in terms of a clause which will find the best of the given list of projects. For example, if we ask:

```
?-best([aa,bb,cc],Answer).
```

it should come back and tell us:

```
Answer=bb,[21.5]
```

since, as the above output shows, bb has the highest expected value of returns.

Exactly the same approach of representing the information in the form of lists is helpful in developing a program for getting revised probabilities using Bayes' Theorem, given the prior probabilities and the conditional probabilities. In the exercises section, I have indicated some considerations which are relevant to designing such a program, and a possible solution in the light of these considerations is given in the solutions section.

EXERCISES

1. In Chapter 4, we saw how the concepts of lists and recursion could be applied for calculating expected values. The same concepts are also very handy for calculating revised probabilities using Bayes' Theorem.

The following suggestions may be helpful to you in designing a program. A suggested solution is given at the end.

(1) Some clauses with the predicate priors would be needed in the knowledge base, e.g. if there are three events with prior probabilities 0.4, 0.1 and 0.5 one could use a clause with the predicate priors with two arguments, the first argument a list of the names of the events, and the second a list of their prior probabilities.

(2) We need to know the conditional probability of some observed indicator, given each of the possible events in (1). One could put in a clause with predicate 'cond', for conditional probability, and three arguments, one the name of the indicator, the next a list of possible events, and the next a list of conditional probabilities.

(3) Recursion would be useful for multiplying each of the prior probabilities by the corresponding conditional probability, and putting the result into a list, until the list of prior probabilities becomes empty. At this stage, the elements of the list would have to be added up, and each element of the list divided by the sum of all of them, to convert them into probabilities. A previous program we have considered, 'addup', would be useful here for adding up the elements.

5

Trees

PREVIEW

So far, we have considered graphs or networks in their most general form, consisting of:

(a) a finite set of nodes
(b) a finite set of arcs between pairs of nodes.

In this chapter, we look at a subclass of graphs known as **trees**. A graph with some additional conditions imposed is a tree, and we we will see what these additional conditions are. We will also look at some useful applications of the concept of a tree.

INTRODUCTION

In the previous chapters, we saw that the basic technique of representing knowledge by using nodes and links between pairs of nodes can be interpreted as meaning different things in different situations or contexts. As a result, the same basic concepts enable us to solve a variety of decision-making problems. What is it that makes this form of representation such a general conceptual tool? We could say that it is based on the fact that, in any situation, knowledge is basically about objects, and relationships between objects. This fits in with the essence of a graph, for the graph consists of relationships between the nodes, and we can interpret what the nodes and the relationships stand for in a variety of ways. The second important feature of knowledge is that, from explicit relationships, we infer or deduce implicit relationships. This is exactly what we do when we trace paths in graphs. A path can be thought of as finding implicit connections between nodes which are not directly connected, by using the explicit links between pairs of nodes which are directly connected. Another way of thinking of a path is to think of it as showing transmission or inheritance of characteristics from some starting point to some ending point. In this chapter, we will look at how this way of looking at what a graph is about is particularly relevant to a special class of graphs known as **rooted trees**.

ROOTED TREES

There are alternative but equivalent ways of defining a rooted tree. We will use a definition which fits in with what we have considered in earlier chapters, i.e. in terms of states and transitions between states.

(a) A rooted tree has an **initial** or **start** state, also called the **root** of the tree. It is also described as constituting level 0 of the tree. In Fig. 5.1 aa is the root state or node.

(b) From the root, transitions are possible to 0 or more states. When there are 1 or more such states, they are called the **successors** of the root. The root is called the **predecessor** of each of these states. An important characteristic of these states is that they have only one predecessor, and this unique predecessor is the root. In other words, there is no transition posssible to any of these states from any state other than the root. These states are collectively said to constitute level 1 of the rooted tree. In Fig. 5.1, bb, cc and dd constitute level 1 of the tree. If the number of successors of the root is 0, then the entire rooted tree consists of just the root. This is not a very interesting or useful example of a rooted tree. But as we will see presently, allowing for the possibility of 0 successors allows us to concisely describe the pattern.

(c) We can complete the description of a rooted tree by saying that each of the nodes or states at level 1 in turn has 0 or more successors which satisfy the same condition as in (b) above, i.e. their only predecessor is a node at level 1. The successors of all the level 1 nodes constitute level 2 of the tree. In Fig. 5.1, ee, ff, gg and hh constitute level 2 of the tree. Each of these in turn can have successors, with unique predecessors in level 2. These collectively constitute level 3 of the tree. In Fig. 5.1 links to level 3 are shown. Similarly, there can be levels 4, 5, 6 . . . .

At any level, a node can have 0 successors, in which case that node is called a terminal node. In Fig. 5.1, dd is a terminal node at level 1, and ff at level 2. Terminal nodes are also sometimes called leaves of the tree.

All the nodes of the tree are referred to as descendants of the root. So, a descendant can be a successor, or a successor of a successor . . . .

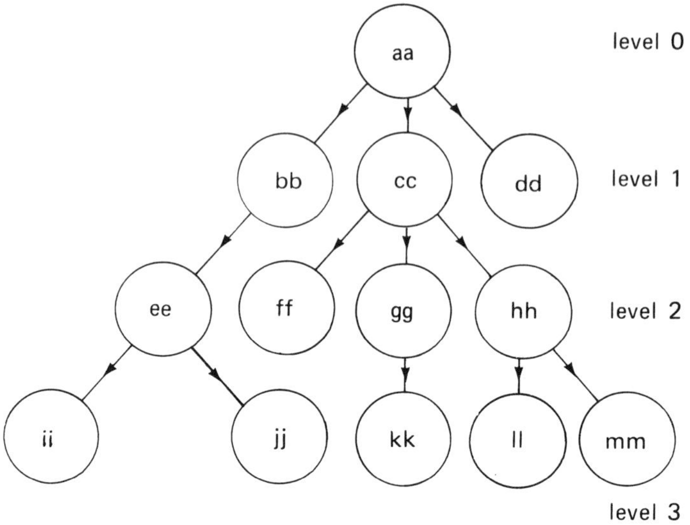

Fig. 5.1.

In an earlier chapter, we discussed the concept of recursive patterns or structures, i.e. patterns which can be described as including subpatterns which are similar to the overall pattern. We can describe a rooted tree recursively as follows.

A rooted tree consists of non-overlapping sets of nodes (i.e. a particular node is contained in exactly one of the sets), called levels 0, 1, 2 . . ., etc., which satisfy the following conditions:

(1) level 0 consists of a single node with no predecessors
(2) any level n, for $n>0$, consists of nodes each of which has a unique predecessor in level $n-1$.

This definition has the characteristic recursive feature that level 6 is described in terms of level 5, which in turn is described in terms of level 4 . . . . The other characteristic of recursive definitions is that ultimately we reach a simple case, and here, when we reach level 0, we describe it as just a single node with no predecessors.

An alternative recursive definition arises from the fact that when we look at any of the level 1 nodes, for example the node cc in Fig. 5.1, we see that it can be regarded as a root with successors ff, gg and hh, which in turn have successors and so on. So, cc is a root of a subtree of the original tree, and the same is true of bb and dd. So we can give an alternative definition of a rooted tree as consisting of a root, and a set of 0 or more successors, each of which is the root of a subtree. The definition of a rooted tree above is clearly an abstraction or generalization of the common device of exhibiting family relationships by using family trees. The levels of the tree are the generations, and the root is a common ancestor from whom relationships are traced. The terminology of family trees is often used in describing trees, by referring to successors and predecessors as children and parents respectively. Also, the predecessor is also referred to as an ancestor, and the the term ancestor is recursively defined by saying:

> a parent is an ancestor.
> any ancestor of a parent is also an ancestor.

From the way a tree is defined, we see that any two nodes have at least one common ancestor, the root at level 0. In addition, they could have 2, 3 . . . common ancestors, depending on whether the subtree to which the two nodes belong stems from a node at level 1 or 2, etc.

The most important implication of the way a tree is defined is that, starting from the root, there is a unique path leading to any node. This follows from the fact that the node in question has a unique parent, which in turn has a unique parent, and by tracing this line of descent backwards, one arrives at the root by tracing the unique path backwards.

One other general concept relating to trees is the idea of a binary tree. A binary tree is one in which there are at most two successors to each node. Sometimes, a binary tree is defined as one which has exactly 2 successors, or 0 successors, but we will use the definition in which 0, 1 or 2 successors are possible.

TREES AND SORTING

A useful way of looking at a tree is to think of it as essentially a way of classifying the nodes. At level 1, we can think of all the nodes in the whole subtree which 'grows' from node cc as a subclass different from the subtree which grows from node bb. Similarly, if we look at level 2 downwards, within the subtree which grows from cc, each of the subtrees which grow from one of the nodes which are the successors of cc is regarded as representing a different category or class. So, a tree represents essentially a hierarchical structure.

An important information processing application of trees, in particular binary trees, is in sorting lists of elements. We can think of a list of names sorted in alphabetical order in the following way. If we pick any member of the sorted list which we call a key or pivot, then it is true that all the elements in the sublist to the left of that element precede the pivot in alphabetical order, and all the elements to the right of the list follow the pivot. If there are some elements equal to the pivot, we can put them either at the end or right of the left list or at the beginning or left of the right list. Now, if we go to the left list, it constitutes a sorted list as well, so it has the same structure: if we pick a pivot from it, it has a left and a right sublist which are all before or after the pivot, and are also sorted. We can keep dividing up these sublists, until we reach a stage where a sublist is empty, in which case there is no more subdivision possible.

We can represent this process of subdivision by a binary tree. Suppose we want to sort a list [p,r,c,s,q,f]. We can think of scanning the list to put all elements which precede p alphabetically in a left group, and all elements which follow p alphabetically in a right subgroup, as in Fig. 5.2.

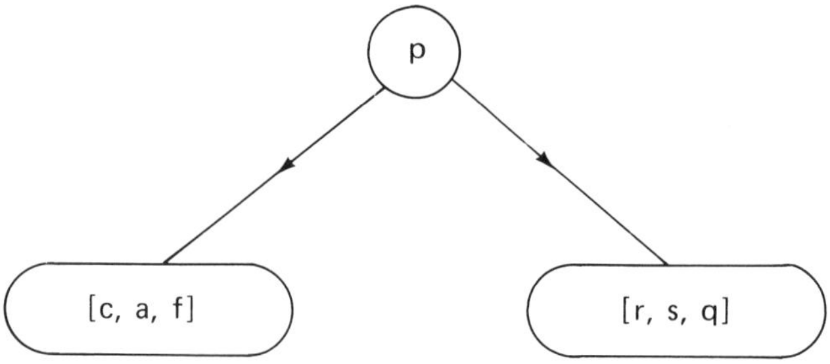

Fig. 5.2.

We can think of repeating the process with these two sublists to take the tree to another level, as in Fig. 5.3.

We see that now the tree shows the correct alphabetical order of the list as a,c,f,p,q,r,s, provided we adopt a certain convention for deciding the order in which the various nodes of a tree should be listed. The convention we have used above can be described as the rule: to list the nodes in a tree, list the nodes in the left subtree first, then the root, and then the nodes in the right subtree.

In this example, it means we must first list the nodes in the left subtree with c as the node. But since this subtree is itself a tree, the rule requires us to first list *its* left subtree, its root, and its right subtree in that order. This leads us to a, c and f. Then we list the root p, and go on to the right subtree with r as node. Here again, we apply the same rule and start by listing the right subtree

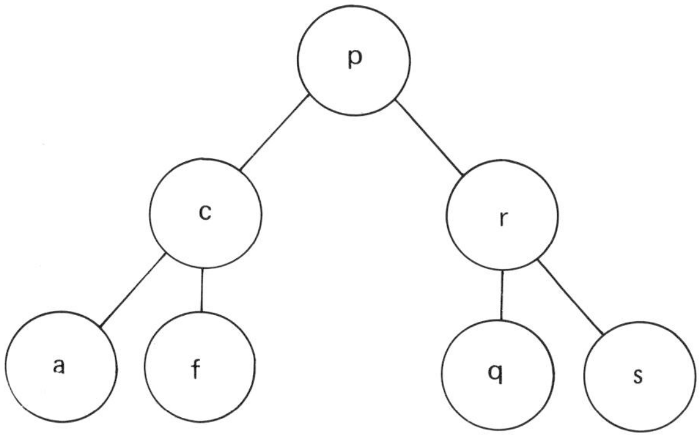

Fig. 5.3.

first, etc., and get the order as q, r and s. This convention for how to list the nodes of a tree is referred to as the **inorder** rule or **symmetric order** rule for listing the nodes. We will look presently at two alternative rules which lead to different ways of listing the nodes of a tree. These methods find an application in other situations. I have given as Appendix 1 to this chapter, a Prolog program which implements this process of representing sorting by dividing the original list into two branches, so that all elements in the left branch are less than the pivot, and all elements in the right branch are greater than or equal to the pivot. Comments have been added before the various clauses to explain their role.

Alternative order for listing nodes

We saw above the unorder or symmetric order rule for listing the nodes in a tree. However, this rule leads to some ambiguity in certain cases. Consider the use of trees for representing arithmetic expressions. We could represent an addition like 3+4 by a binary tree as in Fig. 5.4.

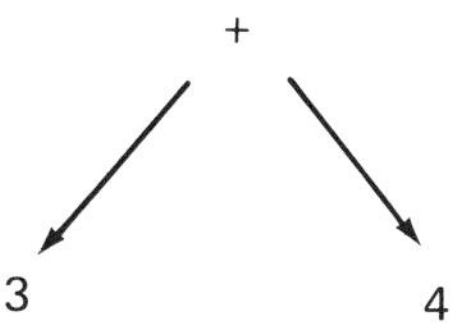

Fig. 5.4.

Similarly a multiplication like 7 * 2 can be represented as in Fig. 5.5.

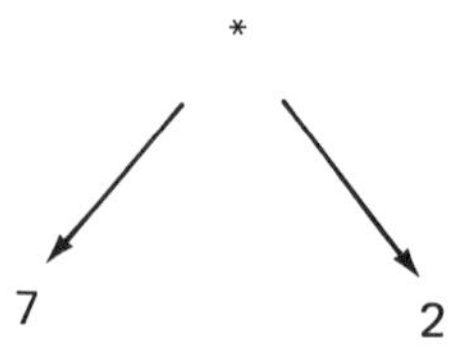

Fig. 5.5.

The node is the operator, and the branches are the two arguments taken by the operator.

Problems arise if the arguments are not just numbers but are expressions, i.e. if we have to go to further levels of the tree. For instance, let us consider the expression 5+3 * 4. This is ambiguous as it stands, because it does not indicate whether it is 5+3=8 multiplied by 4=32, or 5 added to (3 * 4)=17. The ambiguity is resolved by using a convention that multiplication takes precedence, and must be done first, unless parentheses are used to override this convention by writing (5+3) * 4. Using a tree with the inorder convention does not help resolve this ambiguity. Let us use two different trees: Fig. 5.6 for (5+3) * 4, Fig. 5.7 for 5+3 * 4).

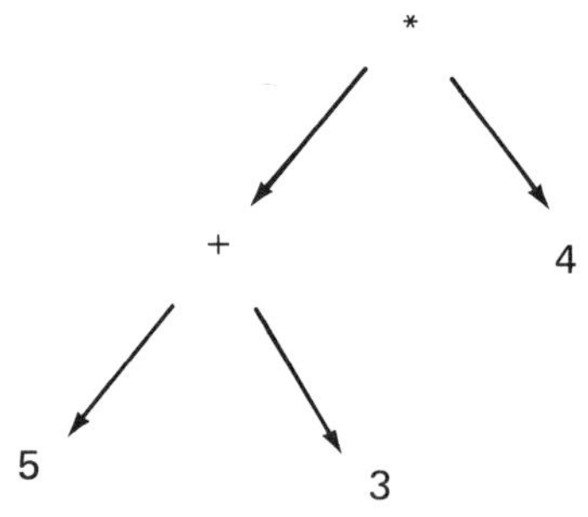

Fig. 5.6.

When we read them in order, we get 5+3 * 4 for both, and still need to use a convention or parentheses to resolve the problem. An alternative way of listing the nodes of a tree helps in this case. This rule is known as the **preorder** rule and is expressed as follows:

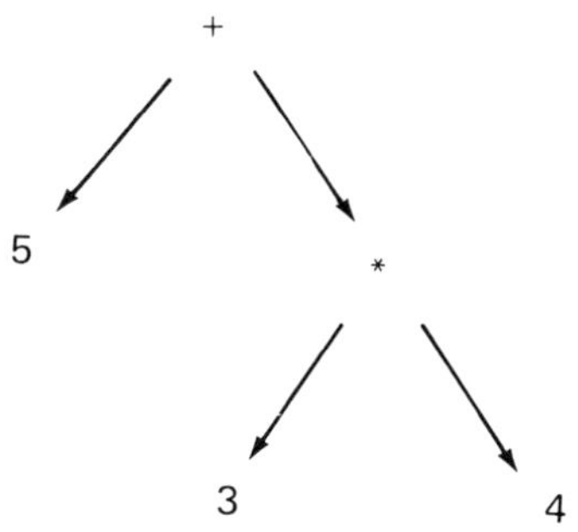

Fig. 5.7.

To list the nodes of a tree, first list the root, then the left subtree and then the right subtree.

When we apply this rule, once again when we go to left or subtree; we apply the same rules to the subtrees. So, in the examples above, the preorder listing of the nodes would lead to the sequences:

*+5 3 4 and
+5 * 3 4

This makes it unambiguous, without any parentheses, that + has to be applied to 5 and 3 and * has to be applied to the result of the + and 4 in the first case, whereas in the second case, * has to be applied to 3 and 4, and + applied to the result and 5. Another illustration of how the preorder rule resolves ambiguity is the following. Let us consider the statement:

A is true AND B is true OR C is true.

This could mean:

(1) A is true
AND either B or C is true
(2) A AND B are both true
OR C is true.

Let us draw a binary tree as in Fig. 5.8.
If we read this using the preorder rule, it becomes:

AND A is true OR B is true C is true

making it clear that two things are joined by AND:

A is true
OR B is true C is true

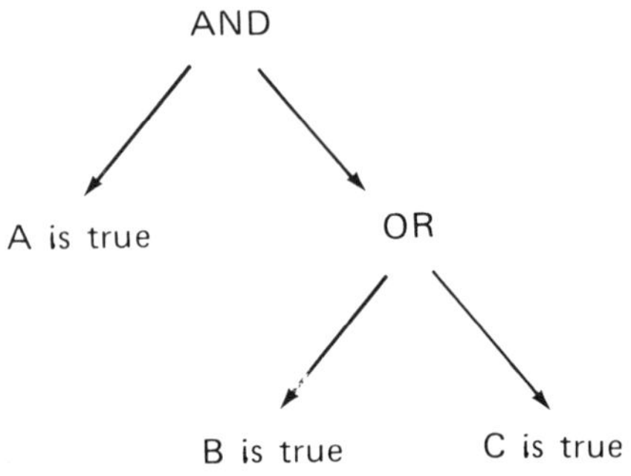

Fig. 5.8.

giving the meaning at (1) above.

To get the meaning at (2) above, a distinctly different tree would be used, as in Fig. 5.9.

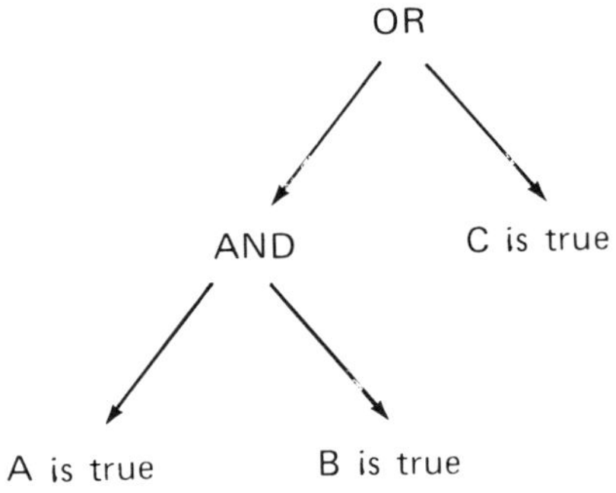

Fig. 5.9.

and we would read this in preorder as:

OR AND A is true B is true C is true.

On the other hand, the inorder reading of both these trees leads to the same ambiguous statement:

A is true AND B is true OR C is true.

TREES AND CODING

To illustrate the flexibility of the idea of a tree as a problem-solving tool, let us look at the application of trees in coding information.

Typically, some code is used to represent a message when it is transmitted. Let us say we have a message which is made of words using five different characters, and we wish to code them into sequences of 0s and 1s for electronic transmission. We can express all possible sequences of 0s and 1s using a binary tree as in Fig. 5.10.

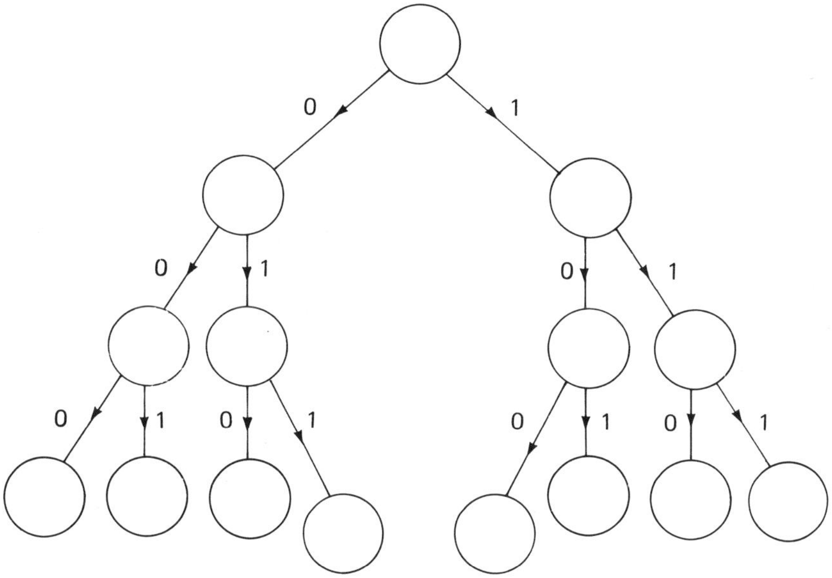

Fig. 5.10.

We can see from this tree that we can have two possible sequences of length 1: 0 and 1. We have four possible sequences of length 2: 00, 01, 10 and 11 and so on. Similarly, since there are two possible choices for the digit at each stage, if we want a four-digit number, there are 2 raised to the power 4 possible sequences. With five different characters in the messages we wish to send, if we want to make them all of equal length, we require at least three digits to make them all different, and on so the average number of 0, 1 characters we use for sending messages will also be three per character in the message. However, if we know that the characters do not all have the same relative frequency, it would be more efficient to use longer sequences for the less frequent characters, and shorter sequences for the more frequent ones. This is the principle used in the Morse code, where the most frequent letter of the English alphabet, e, is transmitted using the shortest sequence of signals, a dot, and the less frequent letters are transmitted using longer patterns of dots and dashes. However, there is one requirement which is very desirable when we use shorter sequences for some characters and longer ones for others. We have to choose the codes in such a way that the code for one of the symbols is not the beginning subsequence or prefix of the code for another. For example, if we use 0 for one character p, and 01 for another character q, because 0 is a prefix of 01, at the receiving end it would not be immediately clear when a 0 is received whether it represents p or is part of the representation of q. In other words, at the receiving end, the characters would not be **instantaneously decodable**, and the messages would have to be held in memory to distinguish between p and q. If we can choose a method in which the condition mentioned above of the code for any symbol not being the prefix of the code for any other symbol is met, we have another advantage in addition to instantaneous decodability. An ambiguous code would be one in which, say, we use 0 for p, 1 for q and 01 for s. In that case, it would not be clear whether a 01 represents the sequence pq or just s. This kind of problem of ambiguity obviously cannot arise when we have a code which satisfies the 'no prefix' rule. It is possible to have codes which are not instantaneously decodable which are not ambiguous either, but among non-ambiguous codes, it is more convenient to have those which are instantaneously decodable.

A method of finding an instantaneous decodable representation which makes the average length as small as possible by assigning longer sequences to less frequent characters is known as Huffman coding and works as follows. Suppose the relative frequencies of five characters we want to code are respectively:

p.	0.3
q	0.25
r	0.25
s	0.1
t	0.1

The Huffman coding method starts by drawing two leaves representing the two least frequent characters, s and t. In this case, we will draw the tree with the leaves at the top, as shown in Fig. 5.11.

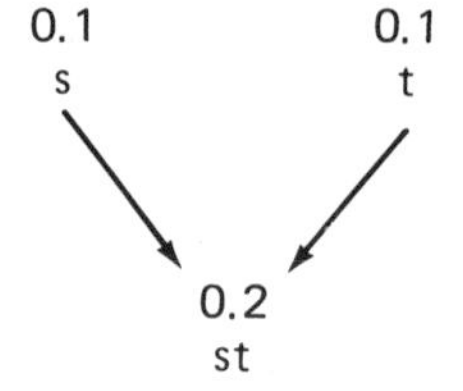

Fig. 5.11.

The logic of the Huffman method is that we can assign 0 and 1 as the last digits for these two characters, to distinguish them. So we can combine them into one, with a relative frequency of 0.1+0.1=0.2 and treat them as a single entity. So, we now have only four entitities to distinguish, p, q, r and st. We apply the same method of combining the two least frequent of these, r and st, into one entity, to be distinguished later by using a 0 for one of them and a 1 for the other, as in Fig. 5.12.

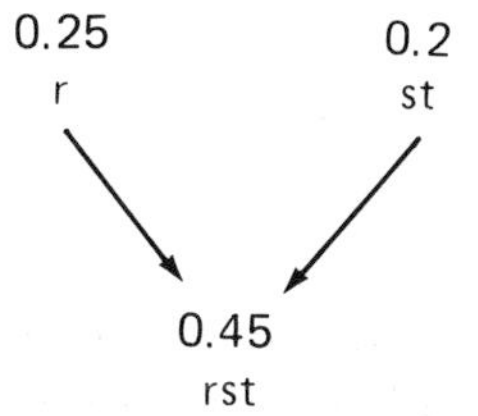

Fig. 5.12.

We repeat this process. Now the two entities with the lowest frequencies are p and q and we combine them in Fig. 5.13

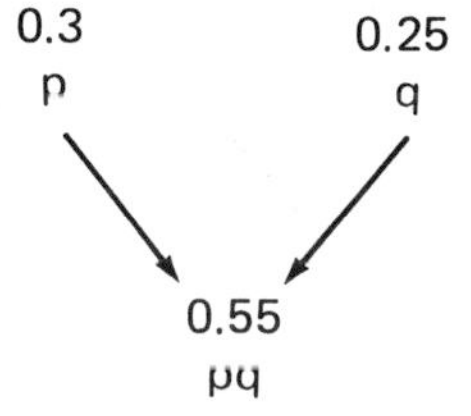

Fig. 5.13.

We now have only two entities, pq and rst, and we can now assign 0 to pq and 1 to rst as the first digit, as shown in Fig. 5.14.

Fig. 5.14.

But we want to distinguish p and q, so we assign 00 to p and 01 to q, as shown in Fig. 5.15.

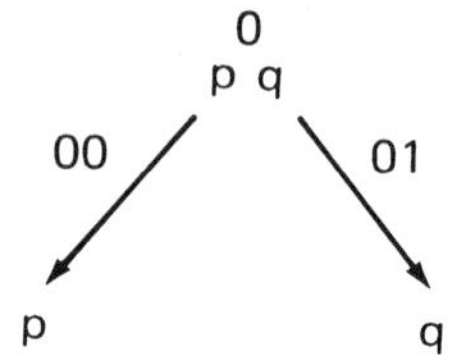

Fig. 5.15.

Again, we want to distinguish r and st, so we assign 10 to r and 11 to st. And finally, we distinguish s and t by assigning 110 to s and 111 to t, as shown in Fig. 5.16.

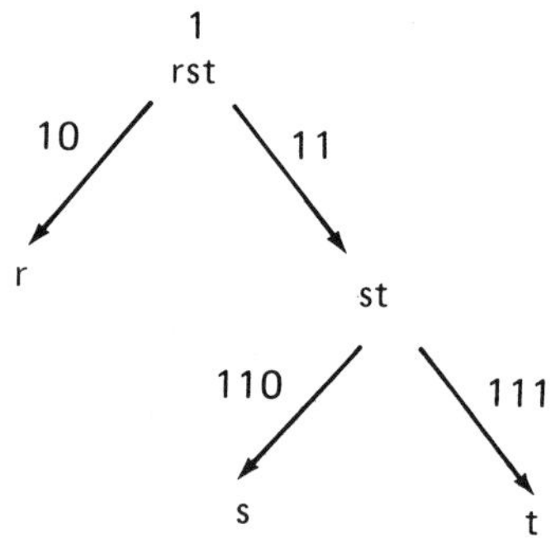

Fig. 5.16.

So, we end up with the coding system:

	Relative frequency	
p	0.3	00
q	0.25	01
r	0.25	110
s	0.1	1110
t	0.1	1111

We see this does ensure that no code is a prefix of another code. Also, the average number of 0, 1 symbols used per character is: $0.3\times2+0.25\times2+0.25\times3+0.1\times4+0.1\times4 = 2.65$. So, we gain by using this variable length coding when the characters have different relative frequencies.

MORE GENERAL TREES

We started by defining a tree as starting with a specific node, the **root** of the tree, and then going on to successive levels of descendants. There was a special feature that a node at any level could have exactly one predecessor at the previous level, though it could have any number of successors at the next level. There was a specific direction associated with the levels, going from predecessors to successors. So, what we looked at was not just a rooted tree, it was a *directed* rooted tree. We also saw that a rooted tree can be thought of as consisting of a number of subtrees.

Because each node has a unique predecessor, it also means that each node has a unique 'line of descent' from the root of the tree. Further, if we take any two nodes which are part of the same subroot of the tree, each of them has a unique line of descent not only from the root of the whole tree, but also from the subroot. For example, in Fig. 5.1, kk and mm each has a unique line of descent from aa. But in addition, they both have unique lines of descent from cc, the root of the subtree to which both the nodes belong. A consequence of this fact is that, if we ignore the direction of the links, we could say that there is a unique path from any node in the tree to any other node. This unique path may go through the root aa of the tree, as in the case of the nodes jj and gg. The unique path connecting them is jj–ee–bb–aa–cc–gg. On the other hand, in the case of the nodes kk and ll, the unique path connecting them does not include aa. The path is kk–gg–cc–hh–ll. But the path does pass through cc, which is the root of the subtree to which both kk and ll belong.

When we generalize from a directed rooted tree and ignore the direction of the links, we could redraw the tree without distinguishing any particular node as the root. The characteristic that there is a unique path from any node to any other node is preserved. One way of redrawing the tree would be as in Fig. 5.17. This feature that there is a unique path from any node to any other node has lead to some useful applications of the concept of a tree. We look at a couple of such applications below.

MINIMAL SPANNING TREES

Suppose, in Fig. 5.17, instead of having the arcs as shown in the diagram, we had only the nodes, and were also given the information that we could choose

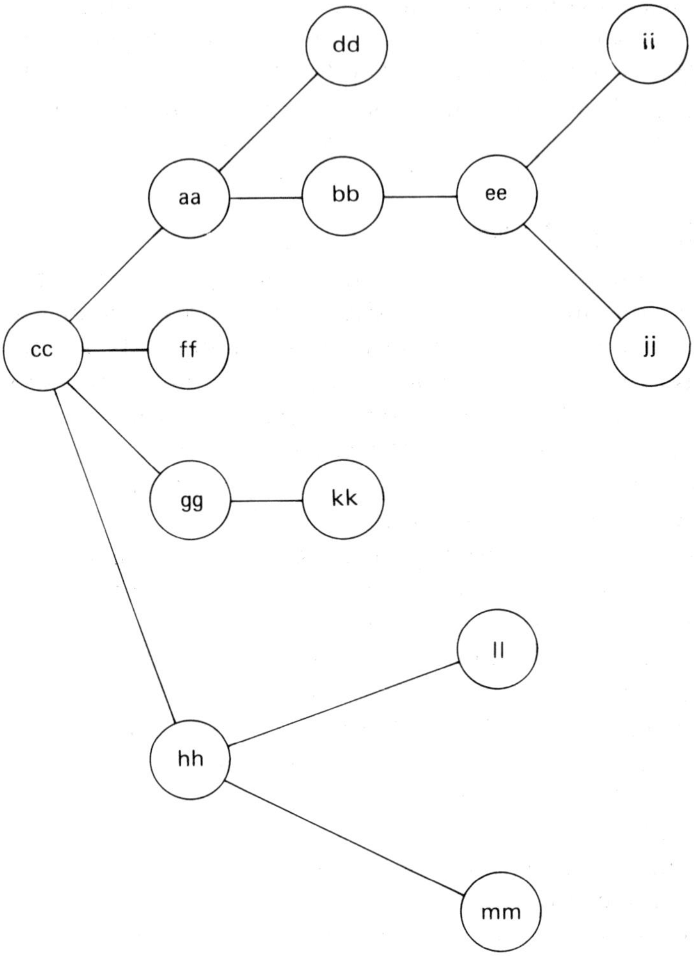

Fig. 5.17.

to have arcs from any node to any other node, with the distances from any node to any other node known. Let us say each node represents a city, and we need to construct a transportation network so that there is a path from any city to any other city. A pattern in which there is a path from any node to any other node is said to **span the whole collection of nodes**. Of course, the extravagant solution would be to construct a link from each city to each other city, with 78 links. (There would be 78 links, because there would be 12 from aa to each of the other cities, 11 from bb to the cities cc, dd . . . , 10 from cc to dd . . . etc., and 12+11+10+. . . 1=78.) But suppose the aim is to provide the paths so that the total distance is minimum. The tree shown in Fig. 5.17 is one possible solution which uses only 12 links. The characteristic of trees which we have seen, that there is a unique path from any node to any other node, is relevant here, for it tells us that a tree pattern would be shorter than any other pattern which has more than one path between two nodes. This follows from the fact that if there is more than one path between certain pairs of nodes, then there must be some redundant link which can be removed without affecting the goal of having a path from any node to any other node. However, the way the tree is drawn in Fig. 5.17 is not the only way a tree can be drawn which spans all 13 nodes. In fact, it can be proved that with n nodes, there are n^{n-2} possible trees. With a large number of nodes, evaluating all possible trees to find the shortest is not a viable procedure. The problem of finding the shortest tree which spans all the nodes turns out to have some simple and elegant solutions, which we look at now.

To illustrate the method, let us look at a problem with five nodes, node_1, node_2, node_3, node_4 and node_5, with distances as given in Table 5.1.

Table 5.1

From	To				
	node_1	node_2	node_3	node_4	node_5
node_1	—	2	3	6	5
node_2	2	—	3	7	4
node_3	3	3	—	2	4
node_4	6	7	2	—	5
node_5	5	4	4	5	—

The distance from any node to itself is shown as '—' in the table, because we are not concerned with building any links from any node to itself.

One of the methods for finding a minimum spanning tree is called **Prim's method** in honour of the person who first put it forward. It works as follows:

STEP 1: Select the shortest of the links. In this example, we would select node_1 to node_2, which has distance of 2. (Actually, here node_3 to node_4 is equally short, so we could equally well start with that.)

At this stage, we have a subtree of the tree we are looking for, with only two nodes connected. As we develop the solution, more and more nodes are connected to this subtree, till it spans all the nodes.

STEP 2(1): Select the shortest link possible which will connect the nodes already connected to a new node not yet connected to them.

In this case, our choice would be node_1 to node_3 with a distance of 3. We do not select node_3 to node_4 with a distance of 2, because it does not connect up with the nodes we already have in the subtree.

STEP 2(2): Repeat the procedure as in Step 2, finding the shortest link from one of the nodes in the subtree to one of the nodes not yet in the subtree.

In this case, our choice would be node_3 to node_4 with a distance of 2.

STEPS 2(3), 2(4)... Keep repeating the process of selecting a link which will connect a node not yet in the subtree to the nodes already in the subtree, till all the nodes are connected.

The number of repetitions of the step depends of course on the number of nodes. Here, we have five nodes, and after selecting two of them to start with, we need only three more steps to connect the other three. So, we finish at step 2(3) by finding the shortest link which will connect node_5, i.e. node_5 to node_2 with a distance of 4. (Alternatively, the choice could be node_5 to node_3, which has also a distance of 4.) The solution found is shown in Fig. 5.18. The key point in this method is that we make sure that each time a new

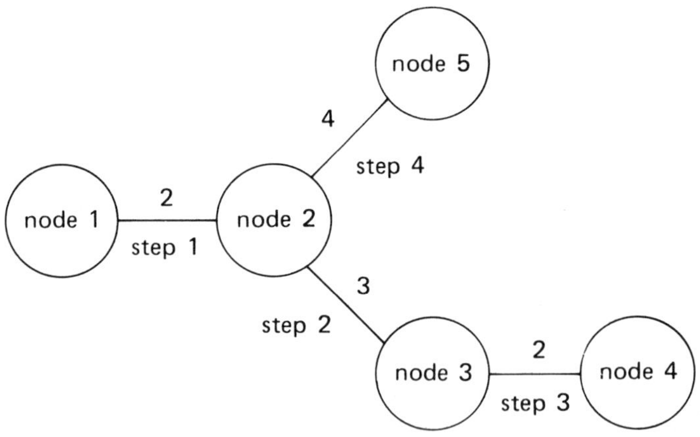

Fig. 5.18.

node is connected to the existing subtree no redundant links which give alternative paths between nodes are introduced.

As our example shows, the minimal spanning tree need not be unique, but it will be not greater than any alternative spanning tree. Incidentally, it is clear that in general, a tree with n nodes would have $n-1$ arcs in it, since starting with any node, we would use one arc to establish connections to each of the other $n-1$ nodes in turn.

A second method of finding a minimal spanning tree is based on selecting the shortest possible link at each stage, whether or not it connects with the nodes already selected, *provided no loops or circuits are introduced*. This method is called **Kruskal's method** after the person who designed it. To see how it would work in the example above, if our first choice as before is node_1 to node_2, then our second choice would be node_3 to node_4 with a distance of 2. So, we do not have a subtree at this stage, but two disconnected portions of our final tree, namely the links node_1 to node_2 and node_3 to node_4. We look again for the shortest possible link we can select, and we can select node_1 to node_3 with a distance of 3. Now, when we repeat this step, the shortest unused link is node_2 to node_3 with a distance of 3 *but we cannot use it because it would create a loop* node_1 to node_2 to node_3 to node_1. So, we go on to look for the next shortest, and select node_2 to node_5, with distance of 4, which will not lead to any loop.

To repeat, the difference between the two methods is that one of them develops a subtree at each stage, whereas the other may select disconnected portions of the final tree, but ends up with a tree. Both would end up with one of the minimal length spanning trees, if there is more than one tree in a particular situation, or with the unique one if there is only one spanning tree of minimal length.

TREES AND REASONING

We have looked at some useful applications of the concept of a tree for representing the knowledge relevant to different types of problems. Now let us look at a different but equally fundamental application, where trees help us to understand the basic processes of reasoning, or applying knowledge to solve problems.

Knowledge and Problem Solving

Let us start by considering the question: what is it we do when we solve problems?

On the face of it, there seems to be quite a difference between the various types of situations in which we talk of problem-solving. There are practical problems, like repairing a computer which has stopped working, or fixing a car which does not start, and so on. There are conceptual problems, like solving an algebraic equation to find the values of the unknown variables x and y. Though these two classes of problems seem different, they do have something in commmon. In both these types, we have enough knowledge to be aware of a clear solution procedure, and we apply it to the specific problem in hand. But we also talk of solving problems in situations where we do not fully know the method for solving the problem, for example when we talk of solving the unemployment problem or the drug problem. This third category seems to be significantly different because we do not really understand it fully, yet in a way it is similar. In all the three categories, what is involved is the application of knowledge about a particular area or domain, whether it is the domain of the electronics of computers, or the domain of automobile engineering, or the domain of the theory of algebraic equations, or economics or whatever. The major difference is that in the third category of problems, not only do we apply existing knowledge, but because of the incompleteness of that knowledge, we have to try to create new knowledge as well. In attempting to solve problems of this class, we use existing knowledge as far as it will go, but because it does not give us complete guidance as in the other cases, we try out tentative hypotheses or conjectures, and to the extent they work out, we add them to existing knowledge.

We have talked above about applying knowledge, but what constitutes knowledge? This is a question which has exercised the minds of thinkers since time immemorial, and it is a matter which has many complex aspects. However, for our purposes we will use a simple operational definition, that is, a definition in terms of the actual actions or operations involved. In practice, when do we say a person knows or has knowledge about a particular matter? We could say that the test is that if a person can answer questions or solve problems relating to certain matters, we say that he or she knows and understands these matters. In fact, probably we can go further and say that solving problems can be thought of as a form of answering questions. For example, knowing how to fix a computer which has broken down is in effect knowing the answer to the question: what should be done to make the computer work again? Alternatively, one can think of answering a question as a form of solving a problem, of supplying missing information, for example in answering the question: what is the capital of the USSR?, the problem of finding some information is being solved. However, it is convenient in some circumstances to think in terms of solving a problem and in other situations in terms of answering questions, so we will think of knowledge, in operational terms, as associated with either or both of these activities.

Applying knowledge

The current approach in computer applications is to think of computer software as operating with two principal components: firstly, a **knowledge base** which contains the relevant knowledge in a particular domain or area of application which is needed for solving the problem — the knowledge that a human being trying to solve the problem would bring to bear on it. Secondly, an *inference engine*. While the name engine suggests some kind of hardware, what this refers to is really the part of the software which uses the knowledge base to draw conclusions. This distinction is one which is useful in considering not just computerized problem-solving, but also the process by which human beings solve problems without computers, the process which we usually refer to as thinking.

The 'inference engine' used by human beings goes by various names, such as logic or, more properly, deductive logic, or the rules of inference. The distinguishing part of the rules of inference is that they give us some general procedure for drawing correct conclusions, given certain specific knowledge about a particular domain. By general procedure, we mean it is not something which applies only to a particular area. For example, an important rule of reasoning or inference is the one which has the historical name of **modus ponens**. This rule says that if we know that

(1) B is true IF A is true
    and we also know that
(2) A is true

we can correctly draw the conclusion that B is true. Here A and B stand for any statements whatever.

These principles governing the process of drawing conclusions or inferences from known facts, using certain valid forms of reasoning, have been sometimes called the 'laws of thought'. This may be too large a claim, since there may be processes other than deductive inference involved in thinking. However, the process of inference or logical reasoning is clearly central to an understanding of knowledge and its uses.

Rules of inference represented by trees

The subject of logic or inference has been studied by thinkers for more than two thousand years. As mentioned earlier, its major concern can be described as the following: what are the valid rules of reasoning, of drawing conclusions based on known or assumed facts? It is important to keep in mind that logic is concerned with valid or reliable forms of reasoning, and that the truth of the conclusions drawn depends on the truth of the assumptions or facts with which one starts. These assumptions are called the premises of the argument. For this reason, the rules of inference are called truth-preserving, that is, if the premises we start with are true, then if we use valid rules of reasoning, we will arrive at true conclusions. Two points are worth noting about this.

Firstly, if we start with false assumptions, using valid forms of reasoning will not somehow lead us to the truth. Secondly, one could arrive at true conclusions even by starting from false assumptions, and/or using invalid forms of reasoning. This could happen if the conclusion happens to be true for some other reasons. The use of valid reasoning only assures us that IF the assumptions we start with are true, and the rules we use are valid, THEN the conclusion must be true.

How do we use a tree to capture the fundamental rule which is the basis of all valid inference, the rule called modus ponens:

IF we know that
(1) IF certain conditions are true
THEN certain conclusions follow.
(Or, as it is sometimes expressed:
Certain conclusions are true
IF certain conditions are true)
AND
(2) IF we know that the asssumptions are true
THEN the conclusions are true.

Let us draw a tree as in Fig. 5.19, with the root labelled as A for the

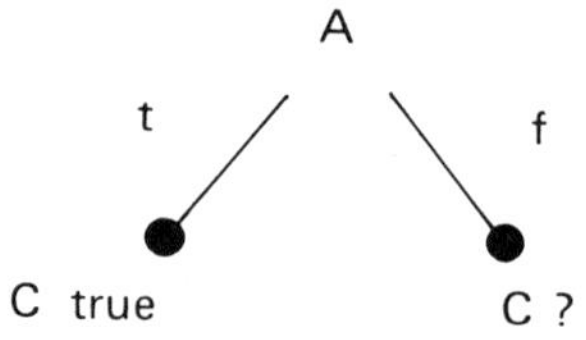

Fig. 5.19.

assumptions we start with, and C for the conclusions. The assumptions we start with are also called the **premises** of the argument. We see that the statement that C is true if A is true is represented by the branch labelled 't' for true, which leads from the premises to the conclusions.

What about the branch labelled 'f' for false? This captures a second important rule in inference which says:

IF we know that
(1) IF A is true
THEN C is true.
AND

(2) A is not true.

where A is some set of premises, and C is the conclusion, it is *not* valid to draw the conclusion that C is not true. If the assumptions A are not true, we cannot conclude that C is false, because it is possible that C is true for some other reason. This is why a question mark is shown on the 'f' branch, to indicate that we are unable to say anything about the truth of the conclusions, when we know the premises are not true.

Another important rule of inference is the one known as **modus tollens**, which says that given that:

(1) IF A is true
THEN C is true.

AND

(2) C is not true.

we can validly conclude that A is not true.

Again, we can see from Fig. 5.19 that this must be so. The t path from A necessarily leads to C, so if C is not true, it means that there is no t path starting from A.

The next important facet of deductive inference is that typically an argument is based not on just a single condition or premise, called a simple statement, but on a combination of such statements. Such a combination is called a compound statement. Secondly, these combinations involve one or more of three key connecting words, AND, OR and NOT, called the logical connectives. The logical connectives are also called **Boolean operators** in honour of the mathematician George Boole. When some simple or compound statement is made, it is called an **assertion** and adding NOT is called the **negation** or denial of a statement, and can be written as A is not true, or NOT(A is true), or notA. The interactions of NOT with AND and OR in relation to compound statements are captured by rules which are known as **De Morgan's Laws**, in honour of the British logician who first pointed them out. The first De Morgan law says that if we have two simple statements A and B:

NOT (A is true AND B is true)
is the same as
A is not true OR B is not true, where the 'or' also includes the case where both statements are not be true.

The second law says that

NOT (A is true OR B is true)

is the same as
A is not true AND B is not true.

To show how trees can be used to understand these laws, we can draw a tree as in Fig. 5.20.

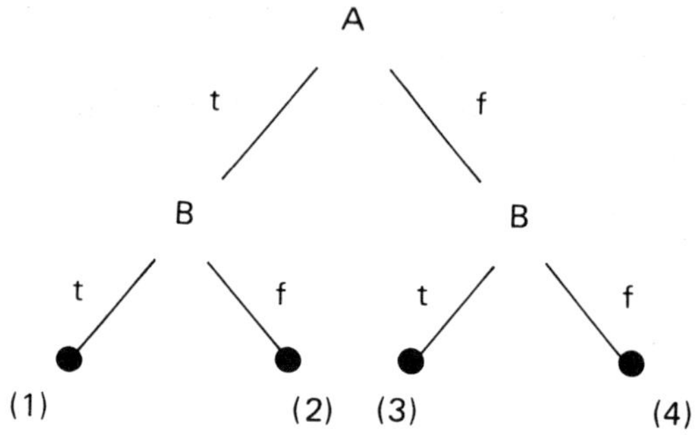

Fig. 5.20.

We can see that the end node (1) results from both A and B being true. So the negation of both being true means we reach any one of nodes (2), (3) or (4). Clearly, this means either A is false, or B is false or both are false, which is De Morgan's first law. Similarly, A is true OR B is true or both are true means we can be at any of the end nodes (1), (2) or (3). So the negation or denial of this must lead to node (4), which represents both the statements being false, and this gives us De Morgan's second law.

We can see from De Morgan's laws that we do not need all three of the logical connectives, NOT, AND and OR. NOT along with any one of the other two would be enough. In fact it is possible to simplify this even further by introducing just one logical connective NAND, standing for NOT AND. We can derive AND, OR and NOT from it as follows:

NAND(A,A) meaning NOT (A is true AND A is true)
clearly means A is not true, so it is in fact the same as NOT A.

We can see this by looking at Fig. 5.21. Nodes (2) and (3) represent impossibilities, since A cannot be true and false at the same time. So, the denial of (1), which represents A is true AND A is true must be node (4) representing A is false.

Similarly, to express the third logical connective OR in terms of NAND, we know from De Morgan's laws that:

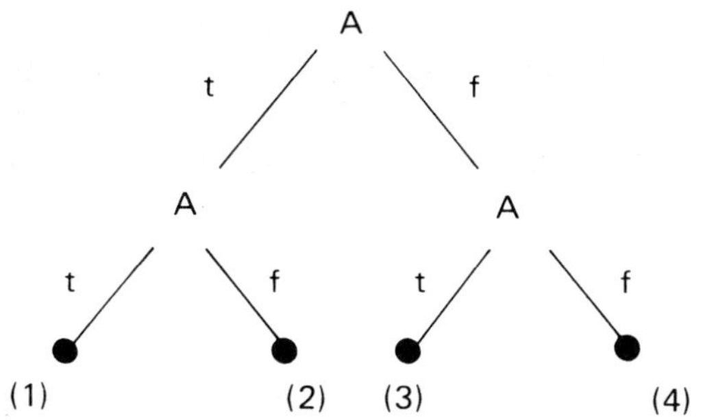

Fig. 5.21.

NAND(A,B) is the same as
notA OR notB.

So, NAND(notA,notB) is the same as
not(notA) OR not(notB) which of course is the same as A OR B.

The practical significance of the fact that all three logical connectives can be expressed in terms of a single operator called NAND arises in its applications in designing electronic circuits. Digital electronic circuits used not just in computers but in various other applications as well require certain switches to operate when one or the other of two input lines is active. These are naturally called OR gates, to which the input lines are connected. Similarly, there are certain switches which are needed to operate only when both input lines are active, the AND gates. Finally, there are circuits where the output has to be opposite to the input, i.e. when the input is active, the output should be off, and when the input is off, the output should be on. The typical symbols used for these circuits are shown in Fig. 5.22.

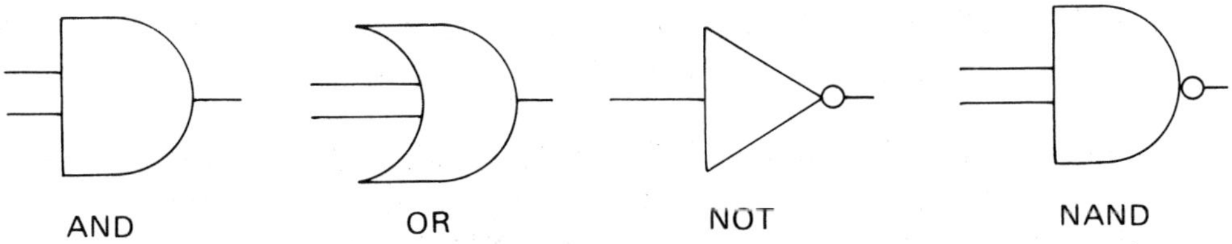

Fig. 5.22.

Because of the logical analysis which shows us that in fact we do not require three types of electronic switches for carrying out these functions, it is possible to take advantage of this fact and design circuits for all functions using just one type of standard NAND switches.

CONCLUSION

In this chapter, we looked at some of the applications of the concept of trees, which are special types of graphs or networks. Trees turn out to be one of the most useful types of graph, and this chapter was meant to convey the feel of this by showing a variety of uses.

APPENDIX

```
/* program sort1.pro */
/* start with boundary case, when list is empty, the sorted list is the empty list
itself */

        sort1([ ],[ ]).

/* for non-empty lists, invoke an auxiliary program 'branch' whose purpose
is to build up the left and right branches. Initially, these are empty lists */

        sort1(List,Result) :-
            branch(List,Result,[ ],[ ]).

/* if there are at least two elements in the list, the branch predicates compare
the first two elements in the list, H1 and H2, and if H2 is greater than or equal
to H1, put it at head of the right branch, otherwise put it at head of the left
branch. After this, the program recurses with the rest of the list other than
H2 */
/* in the version of Prolog used for this program, for numbers >= etc., are
used, and for characters, @>= etc. */

        branch([H1,H2|Rest],R,Left,Right):-
            H2@>=H1,
            branch([H1|Rest],R,Left,[H2|Right]).
        branch([H1,H2|Rest],R,Left,Right):-
            H2@<H1,
            branch([H1|Rest],R,[H2|Left],Right).
```

/* when the list is down to one element, the left and right branches have to be sorted. These sorted lists are then appended together with the pivot element H1 in the middle */

```
branch([H1],Result,Left,Right):-
        sort1(Left,A), sort1(Right,B),
    append(B,[H1|A],Result).
    append([ ],Any,Any).
        append([H|T],Sec,[H|Rest]):-
            append(T,Sec,Rest).
```

The program is invoked by asking questions like:

```
?- sort1([p,a,j,d],What).
```

The response on the screen is:

```
What=[p,j,d,a]->;
no
?- sort1([d,p,a,c,q,s,r],What).
What=[s,r,q,p,d,c,a]->
```

The results are in descending alphabetical order, but one could use the 'reverse' program considered in an earlier chapter to get them in ascending order if necessary.

6

Overview and further reading

In the introduction, I commented on the 'unreasonable effectiveness' of graph theory. The key ideas seem so simple:

(a) We focus on the fact that the knowledge needed for problem-solving and decision-making can be thought of as having three key aspects:
   (1) the collection of objects in the domain of interest
   (2) some important relationship which is apparent or manifest between some or all pairs of these objects
   (3) certain relationships which are not manifest, but can be derived from the relationships in (2).

(b) We generalize the pattern in (a) and represent the objects as nodes, the manifest relationships as links, and the derived relationships as paths.

(c) Consequently, whatever logical conclusions we are able to derive using the abstract model in (b) translate back into solutions for problems which fit the pattern of (a).

This of course is how all conceptual modelling works. In the preceding chapters we have seen very different domains like solving puzzles or coding messages or multistage decision processes all fall within the scope of the graph theory model. The sample we have seen in this book is a very selective sample of the vast areas in which graph theory has found a role. These include such areas as the construction of compilers and interpreters for programming languages, the relationship of graph theory to linear algebra, the representation of knowledge in artificial intelligence and knowledge based systems, the use of graphs in natural language processing, computer vision applications, etc. However, if one has a firm grip on the fundamentals of the theory, as set out in the examples we have discussed in this book, it should be possible to make a smooth transition to the application of graph theory in any area.

I give below a short bibliography. There are many, many excellent books to choose from, so, in order not to make the list forbiddingly long, I have selected half a dozen or so which I enjoyed most. I hope the reader's taste and mine will coincide.

FURTHER READING

Books relevant to all chapters in this book

Biggs, Norman L.: *Discrete Mathematics* (Oxford University Press, 1989).
Quite a large section on graph theory.

Busacker, R. G. and Saaty, T. L.: *Finite Graphs and Networks* (McGraw Hill, 1965).
A splendid book, with a lot of applications.

Gondrian, M. and Minoux, M.: *Graphs and Algorithms* (John Wiley, 1984).
Very comprehensive from a theoretical point of view.

Konig, D: *Theory of Finite and Infinite Graphs* (Birkhauser Boston, 1990).
In his introduction to the 1990 English translation of this classic published in 1935, Dr Tutte points out that this book is justly celebrated as the first textbook on the subject, and as one which led to the study of graph theory as a subject in its own right.

Smith, David K. *Network Optimization Practice* (Ellis Horwood, 1982).
Graph theory, and in parallel, programs in BASIC and Pascal.

Wilson, Robin J.: *Introduction to Graph Theory* (Longman, 1985).
As clear and lucid as it is concise.

Book on Prolog

Bratko, Ivan: *Prolog Programming for Artificial Intelligence* (Addison Wesley, 1990).
Extremely readable, takes one smoothly from the basics right up to very complex applications.

Book on probability theory, relevant to Chapter 4

Allen, Arnold O.: *Probability, Statistics and Queueing Theory: With Computer Science Applications* (Academic Press, 1979).

Book on logic, relevant to Chapter 5

Salmon, M. H.: *Introduction to Logic and Critical Thinking* (Harcourt Brace, 1989).

Solutions for exercises

CHAPTER 1

1. It would make no difference whether we say Same and Same, Start and Start or End and End — what we want to convey is that the two variables should stand for the same object.

2. A recursive definition would be to say that the last element in a list is the last element in the tail of the list. The boundary condition would be that when it is the only element in the list, it is the last element.

In Prolog, the boundary condition would be:

```
last(Any,[Any|[]],Any).
```

or alternatively:

```
last([Any],Any).
```

The recursive case would be:

```
last([Head|Tail],Answer) :-
                    last(Tail,Answer).
```

CHAPTER 2

One way of making the jugs program more general would be to modify the pour clauses. Suppose we modify the clause for pouring from jug_a to jug_b as below:

```
pour([Q,jug_a,to,jug_b],[A,B,C],[A1,B1,C]) :-
                              capacity(jug_b,Bcap),
                                Free is Bcap-B,
                              lesser(A,Free,Q),
                                Q>0,
                                A1 is A -Q,
                                B1 is B+Q.
```

Now,we add a clause to the knowledge base as follows:

```
capacity(jug_b,5).
```

We can see that the pour clause will operate in the same way as if we had explicity put in Free is 5−B. But the advantage is that by changing capacity-(jug_b,5). to some other capacity say capacity(jug_b,6). or whatever, one can use the program for a different sized jug jug_b. Similarly, one can add capacity(jug_a,8). and capacity(jug_c,3). to the knowledge base, and modify the other five pour clauses as well, to enable easy changes in capacities of the three jugs.

CHAPTER 3

1. Firstly, let us consider what changes we need the clauses for 'best'. We had the following clauses in the program in the body of the chapter:

```
                    /*        BEST CLAUSE1         */
best([H1|T1],[H2|T2],Finish,Answer,Choices):-
                      sp(H1,Finish,L1),
                      D1 is H2+L1,
                best(T1,T2,Finish,Answer,[D1|Choices]).
/*        BEST CLAUSE2        */
best([],[],Finish,Answer,Choices) :-

                  smallest(Choices,Answer).
```

As mentioned in the guidelines, we saw that the best clause finds the shortest path through each of the nodes in the set of immediate successors, and then puts in the length of that path into the list Choices, by means of [D1|Choices]. What we need to do now is to insert not only the length of the path (D1) but the actual path as well. To do this we modify the 'best' clauses into 'bestnew' clauses as follows:

```
/*  MODIFIED BEST CLAUSES TO FIND THE ACTUAL SHORTEST PATH
AS WELL AS ITS LENGTH   */
```

```
/*        BESTNEW CLAUSE      1        */

 bestnew([H1|T1],[H2|T2],Finish,Answer,Choices):-
                          sp(H1,Finish,[Lengthr|Router]),
                           D1 is H2+Lengthr,
    bestnew(T1,T2,Finish,Answer,[[D1,H1|Router]|Choices]).

/*        BESTNEW CLAUSE          2 */
bestnew([],[],Finish,Answer,Choices) :-
                        smallest(Choices,Answer).
```

The point to note is that when the bestnew clause recursively calls itself in bestnew clause 1, it has: [D1,H1|Router]|Choices] as the argument, instead of the previous[D1|Choices].In other words, it is carrying along not just the distance, D1, but also the immediate successor, H1, and the shortest path found from that immediate successor, i.e. Router. No change is required in the second clause — bestnew clause 2 is the same as best clause 2.

Secondly, because the list built up in Choices has additional elements, it means the 'smallest' clauses must also be modified. These modifications would be as follows:

```
                /*      OLD CLAUSES FOR SMALLEST         */

    smallest([Only],Only).

    smallest([H1,H2|T],Result):-
         H1>=H2,
           smallest([H2|T],Result).

smallest([H1,H2|T],Result):-
         H1<H2,
           smallest([H1|T],Result).

/*  NEW CLAUSES FOR SMALLEST WITH PREDICATE SMALLNEW    */

smallnew([Only],Only).

smallnew([[H1|T1],[H2|T2]|T],Result):-
                 H1>=H2,
                   smallnew([[H2|T2]|T],Result).
smallnew([[H1|T1],[H2|T2|T],Result):-

                 H1<H2,
                   smallnew([[H1|T1]|T],Result).
```

In the old clauses, we had a list of lengths, [H1,H2|T], representing the lengths of alternative paths, and out of the elements H1 and H2, we wanted to carry along only whichever was the smaller of the two, and so in the recursive call we used [H1|T] or [H2|T]. But now, we have a list of lists, the first two elements of which are [H1|T1] and [H2|T2] and having found the smaller of H1 and H2, we want to carry along the path which goes with it as well, so we recurse with [[H1|T1]|T] or [[H2|T2]|T] as the case may be.

Finally, what is the effect of these changes on the clauses for 'sp' ? The old clauses were:

```
/* OLD CLAUSES FOR FINDING JUST THE LENGTH OF THE
                          SHORTEST PATH */
sp(Start,Finish,Length):-
                          follow(Start,[Finish],[Length]).

sp(Start,Finish,Length) :-
                          follow(Start,Allsucc,Distances),
                          best(Allsucc,Distances,Finish,Length,[]).

/* NEW CLAUSES FOR SHOWING ACTUAL SHORTEST PATH
AS WELL AS ITS LENGTH*/

/* bellman2.pro finds shortest path anywhere to dest, and also display route
— this is additional feature over bellman1.pro */

/*   SPNEW CLAUSE1   */

spnew(Start,Finish,[Length,Start,to,Finish]):-
                          follow(Start,[Finish],[Length]).

/*   SPNEWCLAUSE 2   */
spnew(Start,Finish,[Length,Start,to|Rest]) :-
                          follow(Start,Allsucc,Distances),
bestnew(Allsucc,Distances,Finish,[Length|Rest],[]).
```

Looking at clause 2 first, we see that the major change is that when calling 'bestnew' we need to find out not just Length, but also the path that goes with it, so we use [Length|Rest], the Rest being for finding the acutal path. Also, in the head of the clause, we need just the length, so we put in [Length,start,-to|Rest]. so that we will get a response giving the path as well as in the sample output below . The same change is needed in spnew clause 1 as well.

If we put all these modifications into a revised program bellman2.pro and invoke it,we get the responses as below, which gives us the actual paths in each case:

```
?- spnew(From,dest,Answer).

From=dd
Answer=[5,dd,to,dest]->;
```

```
From=ee
Answer=[4,ee,to,dest]->;

From=start
Answer=[13,start,to,bb,bb,to,ee,ee,to,dest]->;

From=aa
Answer=[9,aa,to,dd,dd,to,dest]->;

From=bb
Answer=[9,bb,to,ee,ee,to,dest]->;

From=cc
Answer=[7,cc,to,ee,ee,to,dest]->;
no
?-
```

2. In the light of the considerations set out in the exercises section, the sp program could be modified as below to find the shortest path from start to any node:

```
/*          listing for bellman3.pro        */
/*     finds paths and displays routes from start to anywhere       */

sp(Start,Finish,[Length,Start,to,Finish]):-
                        before([Start],Finish,[Length]).
                sp(Start,finish,[Length|Back]):-
                        before(Allpred,Finish,Distances),
                best(Start,Allpred,Distances,[Length,_|Rest],[]),
                        append(Rest,[to,Finish],Back).
append([],Second,Second).
append([H|T],Second,[H|Rest]):-
                        append(T,Second,Rest).

best(Start,[H1|T1],[H2|T2],Answer,Choices):-
                        sp(Start,H1,[Lengthr|Router]),
                        D1 is H2+Lengthr,
            best(Start,T1,T2,Answer,[[D1,H1|Router]|Choices]).


best(Start,[],[],Answer,Choices) :-
                        smallest(Choices,Answer).


smallest([Only],Only).

smallest([[H1|T1],[H2|T2]|T],Result):-
                H1>=H2,
                   smallest([[H2|T2]|T],Result).

smallest([[H1|T1],[H2|T2]|T],Result):-
```

```
                H1<H2,
                  smallest([[H1|T1]|T],Result).

before([start],aa,[5]).
before([start],bb,[4]).
before([start],cc,[7]).

before([aa,bb],dd,[4,6]).
before([bb,cc],ee,[5,3]).
before([dd,ee],dest,[5,4]).
/*        end of listing of bellman3.pro        */
```

Invoking the program with ?-sp(start,To,Route). gives the output:

```
?- sp(start,To,Route).
 To=aa
Route=[5,start,to,aa]->;

To=bb
Route=[4,start,to,bb]->;
To=cc
Route=[7,start,to,cc]->;

To=dd
Route=[9,start,to,aa,to,dd]->;

To=ee
Route=[9,start,to,bb,to,ee]->;

To=dest
Route=[13,start,to,bb,to,ee,to,dest]->;

no
```

3. The second and third clauses for smallest work by picking whichever is the smaller of the first two elements in the list, H1 and H2, and then recursing after dropping the larger element. For example in smallest clause 2:

```
smallest([H1,H2|T],Result):-
               H1>=H2,
                 smallest([H2|T],Result).
```

we see that if H1>=H2, H1 is dropped and the recursion proceeds only with H2 and the rest of thc list, T. So, to make it drop the smaller of the two, the clause could be modified as:

```
/*        LARGEST CLAUSE        2  */
largest([H1,H2|T],Result) :-
```

```
                H1>=H2,
              largest([H1|T],Result).
```

On the same lines, the largest clause 3 would be:

```
/*      LARGEST CLAUSE     2   */
        largest([H1,H2|T],Result):-
              H1<H2,
                largest([H2|T],Result).
```

so that the smaller of the two is again dropped. When finally only one element is left, we need to pick that:

```
/*      LARGEST CLAUSE 1        */
        largest([Only],Only).
```

With the modification above, the rest of the program is as below, and you will see it is practically the same as the program for the shortest route:

```
/* KNAP1.PRO SOLVES KNAPSACK PROBLEM BY FINDING LON-
GEST PATH  */
knap1(Start,Finish,Length):-
                  follow(Start,[Finish],[Length]).

knap1(Start,Finish,Length) :-
                follow(Start,Allsucc,Distances),
                best(Allsucc,Distances,Finish,Length,[]).

best([H1|T1],[H2|T2],Finish,Answer,Choices):-
                  knap1(H1,Finish,L1),
                   D1 is H2+L1,
              best(T1,T2,Finish,Answer,[D1|Choices]).

best([],[],Finish,Answer,Choices) :-
                 largest(Choices,Answer).
```

CHAPTER 4

Based on the considerations given in the exercises section, the following is a possible solution:

```
/* prior and conditional probabilities put in knowledge base  */

     priors([aa,bb,cc],[0.4,0.1,0.5]).
     cond(ee,[aa,bb,cc],[0.8,0.6,0.4]).
     cond(ff,[aa,bb,cc],[0.5,0.5,0]).
```

```
cond(gg,[aa,bb,cc],[0,0,1]).

cond(predstable,[stable,worsen],[0.8,0.1]).
priors([stable,worsen],[0.6,0.4]).
```

/* one would invoke the program with a query :

```
?-bayes(aa,bb,cc],ee, Answer).
```

or

```
?-bayes([stable,worsen],predstable,Answer).
```

The first argument is the list of possibilities, the second the observed event or forecast, and the third the answer we want */

```
bayes(Possibilities,Event,[Possibilities|Answer]) :-
            priors(Possibilities,Priorprobs),
            cond(Event,Possibilities,Condprobs),
            revise(Priorprobs,Condprobs,Answerr,[]),
              reverse(Answerr,Answer).
```

/* the prior probabilities and the conditional probabilities are found from the knowledge base, and using these, 'revise' is called, e.g. as revise([0.4,0.1,0.5],[0.8,0.6,0.4],Answerr,[]). As we will see from the revise clauses, the list of revised probabilities is built up in reverse order, so we need to reverse the list, using a previously defined reverse program. */

```
reverse(Given,Answer) :-
            reversea(Given,Answer,[]).

 reversea([HT],Answer,Carrier) :-
            reversea(T,Answer,[H|Carrier]).

reversea([],Answer,Answer).
```

/* revise clause 1
this successively multiplies the first prior probability by the first conditional probability, then the second . . . until the list is empty. In this example, it will go on till:

```
revise([],[],Answer,[0.32,0.06,0.20]).
```

At this stage, since [] cannot be split into a head and a tail, it will go to revise clause 2 */

```
revise([H1|T1],[H2|T2],Answer,Carrier) :-
                R1 is H1*H2,
                  revise(T1,T2,Answer,[R1|Carrier]).
```

/* revise clause 2 adds up the numbers, using the previously defined addup program, and calls normalize, whose role is to divide each of the numbers by the

```
sum. In the example it will call normalize with normalize([0.32,0.06,0.20],
0.58,Answer) */

revise([],[],Answer,Sequence) :-
                    addup(Sequence,Sum),
                   normalize(Sequence,Sum,Answer).

addup([],0).
addup([H|T],Answer) :-
            addup(T,Restsum),
            Answer is H+Restsum.

/* normalize successively divides each number in the list by the sum of the
numbers*/

 normalize([H1|T1],Sum,[N1|Rest]) :-
                N1 is H1/Sum,
                  normalize(T1,Sum,Rest).
normalize([],_,[]).
```

The output on the screen as a result of invoking the program is as follows:

```
?- bayes([aa,bb,cc],ee,What).
What=[[aa,bb,cc],0.55172414,0.10344828,0.34482759]->;no

?- bayes(Possibilities,Event,What).

Possibilities=[aa,bb,cc]
Event=ee
What=[[aa,bb,cc],0.55172414,0.10344828,0.34482759]->;

Possibilities=[aa,bb,cc]
Event=ff
What=[[aa,bb,cc],0.8,0.2,0.0]->;

Possibilities=[aa,bb,cc]
Event=gg
What=[[aa,bb,cc],0.0,0.0,1.0]->;

Possibilities=[stable,worsen]
Event=predstable
What=[stable,worsen],0.92307692,0.07692308]->;

no
```

Index

ELLIS HORWOOD SERIES IN COMPUTERS AND THEIR APPLICATIONS

Series Editor: IAN CHIVERS, Senior Analyst, The Computer Centre, King's College, London, and formerly Senior Programmer and Analyst, Imperial College of Science and Technology, University of London

Lucas, R.	DATABASE APPLICATIONS USING PROLOG
Lucas, A.	DESKTOP PUBLISHING
Maddix, F.	HUMAN–COMPUTER INTERACTION: Theory and Practice
Maddix, F. & Morgan, G.	SYSTEMS SOFTWARE
Matthews, J.J.	FORTH
Michalewicz, Z.	STATISTICAL AND SCIENTIFIC DATABASES
Millington, D.	SYSTEMS ANALYSIS AND DESIGN FOR COMPUTER APPLICATIONS
Moseley, L.G., Sharp, J.A. & Salenieks, P.	PASCAL IN PRACTICE
Moylan, P.	ASSEMBLY LANGUAGE FOR ENGINEERS
Narayanan, A. & Sharkey, N.E.	AN INTRODUCTION TO LISP
Parrington, N. & Roper, M.	UNDERSTANDING SOFTWARE TESTING
Paterson, A.	OFFICE SYSTEMS
Phillips, C. & Cornelius, B.J.	COMPUTATIONAL NUMERICAL METHODS
Rahtz, S.P.Q.	INFORMATION TECHNOLOGY IN THE HUMANITIES
Ramsden, E.	MICROCOMPUTERS IN EDUCATION 2
Rubin, T.	USER INTERFACE DESIGN FOR COMPUTER SYSTEMS
Rudd, A.S.	PRACTICAL USAGE OF ISPF DIALOG MANAGER
Rudd, A.S.	PRACTICAL USAGE OF REXX
Rudd, A.S.	IMPLEMENTING PRACTICAL DB2 APPLICATIONS
de Saram, H.	PROGRAMMING IN MICRO-PROLOG
Savic, D.	OBJECT-ORIENTED PROGRAMMING WITH SMALLTALK/V
Schirmer, C.	PROGRAMMING IN C FOR UNIX
Schofield, C.F.	OPTIMIZING FORTRAN PROGRAMS
Scott, D.	HUMAN–COMPUTER INTERACTION: A Cognitive Ergonomics Approach
Semmens, L. & Allen, P.	Z FOR SOFTWARE ENGINEERS
Sharp, J.A.	DATA FLOW COMPUTING
Sherif, M.A.	DATABASE PROJECTS
Smith, D. & Sage, M.	EDUCATION AND THE INFORMATION SOCIETY
Smith, J.M & Stutely, R.	SGML
Späth, H.	CLUSTER ANALYSIS ALGORITHMS
Späth, H.	CLUSTER DISSECTION AND ANALYSIS
Stratford-Collins, P.	ADA
Teunissen, W.J. & van den Bos, J.	3D INTERACTIVE COMPUTER GRAPHICS
Tizzard, K.	C FOR PROFESSIONAL PROGRAMMERS
Turner, S.J.	AN INTRODUCTION TO COMPILER DESIGN
Tsuji, T.	OPTIMIZING SCHEMES FOR STRUCTURED PROGRAMMING LANGUAGE PROCESSORS
Wexler, J.	CONCURRENT PROGRAMMING IN OCCAM 2
Whiddett, R.J.	CONCURRENT PROGRAMMING FOR SOFTWARE ENGINEERS
Whiddett, R.J., Berry, R.E., Blair, G.S., Hurley, P.N., Nicol, P.J. & Muir, S.J.	UNIX
Xu, Duan-Zheng	COMPUTER ANALYSIS OF SEQUENTIAL MEDICAL TRIALS
Yannakoudakis, E.J. & Hutton, P.J.	SPEECH SYNTHESIS AND RECOGNITION SYSTEMS
Zech, R.	FORTH FOR THE PROFESSIONAL

Computer Communications and Networking

Currie, W.S.	LANS EXPLAINED
Deasington, R.J.	A PRACTICAL GUIDE TO COMPUTER COMMUNICATIONS AND NETWORKING, 2nd Edition
Deasington, R.J.	X.25 EXPLAINED, 2nd Edition
Henshall, J. & Shaw, S.	OSI EXPLAINED, 2nd Edition
Kauffels, F.-J.	PRACTICAL LANS ANALYSED
Kauffels, F.-J.	PRACTICAL NETWORKS ANALYSED
Kauffels, F.-J.	UNDERSTANDING DATA COMMUNICATIONS
Muftic, S.	SECURITY MECHANISMS FOR COMPUTER NETWORKS